Volume 1

# REGIONS IN RECESSION AND RESURGENCE

# REGIONS IN RECESSION AND RESURGENCE

MICHAEL CHISHOLM

LONDON AND NEW YORK

First published in 1990

This edition first published in 2015
by Routledge
2 Park Square, Milton Park, Abingdon, Oxon, OX14 4RN

and by Routledge
711 Third Avenue, New York, NY 10017

*Routledge is an imprint of the Taylor & Francis Group, an informa business*

*British Library Cataloguing in Publication Data*
A catalogue record for this book is available from the British Library

ISBN: 978-1-138-85764-3 (Set)
eISBN: 978-1-315-71580-3 (Set)
ISBN: 978-1-138-85528-1 (Volume 1)
eISBN: 978-1-315-72041-8 (Volume 1)
Pb ISBN: 978-1-138-85530-4 (Volume 1)

**Publisher's Note**
The publisher has gone to great lengths to ensure the quality of this reprint but points out that some imperfections in the original copies may be apparent.

**Disclaimer**
The publisher has made every effort to trace copyright holders and would welcome correspondence from those they have been unable to trace.

# Regions in recession and resurgence

MICHAEL CHISHOLM

*St Catharine's College, Cambridge*
*Department of Geography, Cambridge*

London
**UNWIN HYMAN**
Boston    Sydney    Wellington

Published by the Academic Division of
**Unwin Hyman Ltd**
15/17 Broadwick Street, London W1V 1FP, UK

Unwin Hyman Inc.
955 Massachusetts Avenue, Cambridge, MA 02139, USA

Allen & Unwin (Australia) Ltd
8 Napier Street, North Sydney, NSW 2060, Australia

Allen & Unwin (New Zealand) Ltd
in association with the Port Nicholson Press Ltd
Compusales Building, 75 Ghuznee Street, Wellington 1, New Zealand

First published in 1990

**British Library Cataloguing in Publication Data**

Chisholm, Michael
   Regions in recession and resurgence
1. Regional economic growth
I. Title
339.5

ISBN 0–04–330062–6
ISBN 0–04–330063–4 pbk

---

**Library of Congress Cataloging in Publication Data**

Chisholm, Michael, 1931-
   Regions in recession and resurgence / Michael
Chisholm.
    p. cm.
Includes bibliographical references.
ISBN 0–04–330062–6. – ISBN 0–04–330063–4 (pbk.)
1. Regional economics. I. Title.
HT388.C48 1990
338.9–dc20

Typeset in 10 on 11 point Times
by Computape (Pickering) Ltd, N. Yorkshire
and printed in Great Britain by
Cambridge University Press

*To Judith*

# *Contents*

# List of tables

# List of figures

# *Preface*

Students approaching problems of regional growth face a bewildering array of ideas and arguments, concerning both theory and practice. To understand the sharp differences in the way regional growth is explained, and the equally clear divergence of policy prescriptions, it is necessary to be aware of the diverse intellectual traditions which are drawn upon. Very often, these traditions are implicit rather than explicit. I have attempted to explain briefly the main relevant strands of economic thought – from neo-classical ideas to supply-side thinking – and to show how these map into divergent schools of thought about regional growth processes. Economic doctrine in general, and regional growth theory in particular, has evolved in the context of major changes in the global economy, and especially the increasingly open nature of national economies. Ideas concerning the reasons for trade and for the sources of national growth have changed quite markedly in recent decades. The changing empirical reality and the evolution of theoretical ideas have both had an impact on the way that regional economic growth is perceived.

Two main themes are evident. First, no school of economic thought – neo-classical, Keynesian, monetarist or whatever – has a monopoly of wisdom. The problem, therefore, is not so much to decide which is right, but to decide which throws the most useful light on which problems. The same conclusion holds for the various theories of regional growth, whether these be based on neo-classical assumptions of free mobility of the factors of production, export multiplier analyses or on entrepreneurship and innovation. In the second place, ideas concerning the sources of growth at the national level and the reasons for international trade have changed over time, to place greater emphasis on man-made attributes (capital, in the form of productive assets, and the skills and abilities of people and firms) relative to the endowment of resources and advantages of location.

In structure, therefore, this book is an extended essay, setting out a viewpoint and an argument, providing an approximate map of a large and complex terrain. The map is imperfect. However, I believe that it will be helpful as a guide, that it shows linkages between global and regional processes in a novel way, and that it will help to show the way in which general economic theory and theory about regional growth interact.

My own thinking has been influenced by the work of many people – too many to recollect them all. I have a particular debt, though, to Bob Bennett for the encouragement which he gave at a crucial time when the ideas for this book were being formulated. Ross Mackay, John Parr and Eric Sheppard were all instrumental in sorting out particular issues, and John McCombie provided valuable comments on the formulation of economic theory. Two close colleagues, David Keeble and Ron Martin, have over the years provided considerable stimulus, and David was kind enough to read a draft of the entire text: his numerous constructive comments were very helpful. I am also much indebted to successive generations of students at Cambridge who have endured my lectures on the topics contained in this book, and also to graduate students at Minneapolis in 1988 who reacted to provocation with lively discussion. The drawings were all done in the Department of Geography, principally by Lois Judge, and Heather Jarman did a magnificent job in typing the final version.

To all of these people, and many more besides, I owe a considerable debt. Nevertheless, they must not be blamed for the errors of omission and of commission which undoubtedly exist – for these, mine alone is the responsibility.

*Cambridge*                                              Michael Chisholm
February 1990

# CHAPTER 1

# *Introduction*

> Much of [the] academic literature on trends in regional development at best gives very limited attention to global economic shifts and at worst either ignores them or takes them as given.
>
> (Browett and Leaver, 1989, p. 33)

The 1950s and especially the 1960s were a golden age for macro-economic management and also for regional theory and policy. At both levels, it seemed that we had sufficient understanding of how economic systems work to be able successfully to intervene in order to achieve desired ends – notably, greater equality of incomes and of opportunities as between individuals and between regions. In the more advanced countries, unemployment appeared to have been banished and output was rising strongly, to give regular increases in the level of real incomes. The social sciences had come, or were coming, of age. Logical positivism was in vogue and seemed, at least in the field of economics, to be justified by the successes which had been achieved in the management of national economies and of the global economic system.

During the last twenty years or thereabouts, economic circumstances have changed dramatically for most countries. The easy confidence that rapid growth can be combined simultaneously with full employment and acceptably low rates of inflation has disappeared. Economists have engaged in a major and continuing debate concerning the macro- and micro-structure of economies and hence the proper approach to intervention by central authorities (see for example, Vane and Caslin, 1987). This wider debate in economics, set against a backcloth of rapid change in national economies, has been reflected in the literature concerned with regional theory and policy. One recent publication comments that: 'The past few years has been a period of significant regional policy upheaval throughout Europe' (Allen *et al.*, 1988), and another carries the title *Regional Policy at the Crossroads* (Albrechts *et al.* (eds), 1989). The doubts and uncertainties thus expressed have been widely articulated in a large and growing literature, much of which has appeared only very recently (Chisholm, 1987; Dunford, 1988; Massey and Allen (eds), 1988; McCombie, 1988a, 1988b; and Vanhove and Klaassen, 1987).

Discussion of economic growth in general, and regional change in particular, reflects the intersection of two broad sets of change. During the last twenty years or thereabouts, the context in which firms take their decisions has evolved quite rapidly. International competition has become more marked, multinational companies have gained in prominence, technological change has been rapid, and manpower resources have become ever more important relative to other location factors. At the same time, management of national economies and of the international economic system has been less successful in maintaining stability than had been the case earlier in the post-war period. In the second place, there has been a sharp debate among economists concerning the way in which national economies operate and hence the nature of, and limitations to, the power of government to control events to achieve desired ends. Indeed, what constitutes desirable ends is itself a matter in question. This debate is often portrayed as a stark contrast between the Keynesian and monetarist schools of thought. Despite the academic and political rhetoric, the contrast between these two schools of thought is less extreme than is commonly believed. In any case, the debate extends to include contributions from the maligned neo-classical tradition and also those of Marxist persuasion. Changes in 'real-world' conditions have undoubtedly had an impact on economic theorising; in turn, theory has a material influence upon the way that the 'facts' are interpreted.

Despite the sizeable recent literature on regional development and policy, nobody has attempted a general review of the changing circumstances which confront firms and regions, and related these changes to shifts in the way that regional growth processes are conceptualized in the competing schools of economic thought. Because this task has not been essayed, except in a preliminary way (Chisholm, 1987), a good deal of confusion exists regarding recent events and how they should be interpreted, and the policy conclusions to be drawn. This uncertainty can be characterized in the following terms. The 1950s and 1960s saw the ascendancy of Keynesian over neo-classical ideas concerning the mechanisms of regional growth and, hence, the basis for policy intervention. The ideas which emerged from this process have been ably described by Richardson (1969, 1973) and form the basis for the work of Armstrong and Taylor (1985), Brown (1972) and many others. However, the Keynesian consensus has been shattered by events since about 1970, and therewith confidence in the policy prescriptions which had been widely accepted. But nothing of equal authority has been offered to account for differing regional fortunes and to point the way toward clear policy prescriptions. Nor do we have a clear articulation of the circumstances in which one kind of approach may be preferable to another.

The present book seeks to describe the major developments in the 'objective' environment in which decision-making relevant to regional growth takes place, and to consider the implications of these changes as seen in the context of shifts in economic doctrine. This is an ambitious task and one that no doubt is imperfectly accomplished. The end result is not a 'new' theory of regional growth, offered as a replacement for that which currently exists. The aim is altogether more modest. If we can recognise the provenance of our ideas and the circumstances in which these ideas were initially articulated, we will be able to judge their appropriateness for new circumstances as they arise. In an important sense, therefore, this book is an history of ideas, but it is not arranged in the form of a conventional chronology.

The argument will proceed in the following way. First, we will review some of the important changes which have occurred in recent years affecting the environment in which firms take their decisions and regional economies operate. It is these changes, and the circumstances which arise in consequence, which any theoretical discussion must note and incorporate. Chapter 2 is then followed by a review of the major issues that have been, and continue to be, debated among economists regarding the way in which national economies operate and the scope that exists to intervene to achieve stated ends. The discussion of these issues in Chapter 3 is entirely non-spatial. The way that these ideas have been translated, or could be translated, into the spatial domain forms the subject matter of subsequent chapters. The main schools of economic thought are treated in a quasi-chronological order, to show their implications for the manner in which regional economies are visualised to work. From these analyses one can derive the policy conclusions which must follow. Furthermore, one can, to some extent, compare actual experience of regional growth with the conclusions derived from theory, though such comparisons provide only partial and incomplete tests of those theories.

The bulk of the discussion will focus on the developed nations of the world. However, although there are marked differences between the more and the less developed nations, there are common features as well. Some of the literature explicitly encompasses countries at all levels of income and development, and in other cases ideas formulated in the Third World context have had a formative effect on thinking about more advanced nations (e.g., Hirschman, 1958; Myrdal, 1957). Therefore, while the primary focus will be on economies such as Britain and the United States, discussion cannot be limited to them alone, and reference will be made on occasion to less developed nations.

For our purpose, the term 'region' will be used in a particular way. Instead of the geographer's concept of the natural region, we will use the term to denote an area within a nation which enjoys certain powers of government, or at least of administration. The absolute and relative size of regions defined in this way vary considerably. The provinces of Canada and the states of Australia and the United States are much larger, and enjoy much greater power, than the *départements* in France or the German *Länder*. These in turn are bigger and more powerful than the British counties and districts. Figure 1.1 illustrates the range in scale which is implicit in working with administrative or governmental units of this kind. California, the most populous state in the United States, has a total population not far short of two-fifths the number of inhabitants of France, West Germany or Great Britain, and a total product which is now the seventh biggest in the world. In addition, the states, provinces or *Länder* which comprise a federation enjoy considerably greater control over their internal affairs than is the case with administrative units in a unitary state such as France or the United Kingdom; such units may be abolished or modified, and their powers changed, by decisions of the national legislature, in a manner which is not possible in a federal state.

Since a region forms part of a national economy, and since a regional government has fewer powers than are possessed by central government, some scholars argue that regional economic analysis cannot be conducted on the same basis as the analysis of national economies. There are several powerful

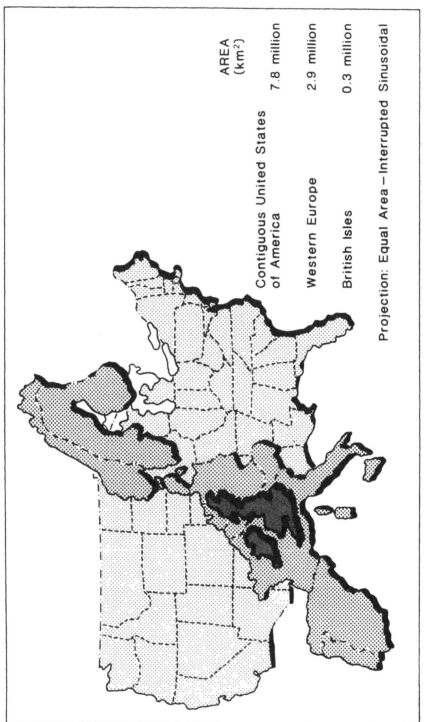

AREA
(km²)

Contiguous United States
of America      7.8 million

Western Europe      2.9 million

British Isles      0.3 million

Projection: Equal Area – Interrupted Sinusoidal

**Figure 1.1**   Comparison of the size of regions

reasons which point in this direction. Regions have no control over the volume of trade (visible and invisible) with other regions, through quotas, tariffs and currency control. They do not usually issue their own currency, though Scotland might appear to be an exception (the £1 note has continued in circulation after its withdrawal elsewhere in the United Kingdom). However, in other respects the currency is identical and there is no possibility of the Scottish pound being revalued against the pound sterling. Because regions share a common national currency, balance of payments surpluses and deficits cannot be compensated by variations in exchange rates, as happens with national economies. Important though these differences may be, much depends on whether one takes a short- or long-run view. Although a national government can set an interest rate or the exchange rate for its currency, in the long run its freedom of manoeuvre is in fact limited. If interest or exchange rates are too high or too low, external forces are apt to generate the requisite countervailing pressures. If one takes a long-run view, the real distinction between regional and national economies is four-fold:

(1) Since the 1920s, more and more nations have taken powers to control immigration. Such controls are now almost universal. In contrast, a region has no effective power to limit the movement of national citizens, either inward or outward.
(2) Regional data, on things such as regional product, trade balances and the like, are, by comparison with national data, either exiguous or non-existent.
(3) The relative power of regional and central government is generally such that whereas regional decisions have a small effect on the nation, a decision taken by central government may have profound effects on one or more regions.
(4) The instruments available to a regional government are generally fewer than those which central government possesses.

The fact that regional data are sparse implies that regional economic analysis must in practice be restricted to fewer dimensions than is possible at the national level.

In an attempt to keep the discussion manageable, we will concentrate attention on the urbanised regions of developed countries, those regions whose economies are dominated by manufacturing industries and/or commercial services. The exclusion of agricultural and resource-based regional economies does not imply that the issues pertaining to their growth and decline are unimportant or simple. Even the restriction to urban-industrial-commercial regions leaves us with an immense canvas, such that the discussion will be incomplete.

One further point must be addressed explicitly. In some of the literature, the distinction is made between growth and development. Growth is thought of as the replication of the existing economic structure – more of the same. Development, on the other hand, is regarded as an expansion of output which is accompanied by changes in the sectoral distribution of activity. To treat growth in this restricted manner is to assume away virtually all the interesting problems of the real world. However useful the distinction may be in certain contexts, it does not seem to be helpful for the present analysis. Therefore, unless otherwise explicitly stipulated, the terms 'growth' and 'development' will be used interchangeably.

# CHAPTER 2

# *The changing context*

There is a tide in the affairs of men,
Which, taken at the flood, leads on to fortune;
Omitted, all the voyage of their life
Is bound in shallows and in miseries.
On such a full sea are we now afloat,
And we must take the current when it serves,
Or lose our ventures.

(Cassius in William Shakespeare's *Julius Caesar*)

In the present chapter, we will examine some of the shifts which have occurred in the structure of the world economy and the ways in which it functions, as an exercise partly of fact and partly of perception. The purpose is to establish some of the framework which is relevant for a consideration of regional growth processes, and as a background against which to view the evolution of theory concerned with these processes.

The fact that the world is an interdependent economic system seems to have been 'rediscovered' in recent years, especially by those inclined to a radical or Marxist view of political economy. Two aspects of this rediscovery are particularly notable. First, the emphasis upon the crisis and instability which observers see in the western (capitalist) economies since the early 1970s. These trends are frequently described in terms of the crisis of Fordism and the rise of new production systems, of the collapse of the Bretton Woods agreement, the associated easement of exchange controls and the rise of multinational corporations. Browett and Leaver (1989) refer to the present-day financial markets as 'unstable', while Lash and Urry (1987) have published a book under the title *The End of Organised Capitalism*. The alleged crisis of Fordism has been explored by Lipietz (1987) among others. The second aspect which has received considerable attention is the proposition that the world economic system is structured in such a way that there is a systematic bias in favour of the 'haves' and against the 'have nots'. This school of thought, variously known as the dependency school or underdevelopment school, is widely associated with the writings of Frank (e.g., 1964, 1966 and 1967). Much of the inspiration was in fact derived from the work of Prebisch at the Economic Commission for Latin America.

However one may interpret the changes, which have undoubtedly been rapid and far-reaching, there is a basic fact about the recent evolution of the world economy which we need to have clearly in mind, namely, the increasing openness of national economies. This increase has occurred in international trade in commodities and services, as well as in the financial markets. The demise of the Bretton Woods agreement in the early 1970s was at least partially due to the very problems of regulation which is implicit in the existence of open, interdependent economies. That which is often interpreted as the instability caused by a change in the mode of operating the international financial markets can as plausibly be regarded as a feature of a dynamic, interdependent world economy which made the former methods of regulation inappropriate. One manifestation has been the unexpected success of Japan as an exporter during the post-war period, and another has been the surprise engendered by the emergence of the four 'little tigers' of Asia – Taiwan, South Korea, Hong Kong and Singapore. More generally, though, the proportion of world output which is traded has been increasing quite rapidly, and this increase has been accompanied by a shift from the 'traditional' pattern whereby developed countries exported manufactures in return for food and raw materials from less developed nations.

## The openness of national economies: commodity trade

There have been some quite remarkable changes in the relative importance of international trade over the last two hundred years or so (Chisholm, 1990). According to the *Times Atlas of World History* (p. 256), the value of world trade (imports plus exports) amounted to 3 per cent of world output in 1800 but had risen to 33 per cent in 1913. Some 71 years later, in 1984, the trade proportion was very similar – 30.5 per cent (GATT, 1986, p. 139; World Bank, 1986, p. 155). It appears that most of the nineteenth-century increase in trade proportion occurred prior to about 1870 or 1880, the proportion thereafter remaining fairly stable until the First World War. Between the two world wars, especially in the 1930s, nations sought to protect themselves against recession by limiting their involvement in trade, and in any case the system of international payments was in considerable disarray following Britain's departure from the gold standard in 1931. The proportion of world output which was traded fell substantially. Since the Second World War steps have been taken to liberalize trade, and the proportion has climbed back to the level which existed in 1913. The rapidity of this climb between 1970 and 1980 is shown in Table 2.1.

During the nineteenth century, international trade was dominated by primary products such as grain and other foodstuffs, textile fibres and timber (three-fifths). The other two-fifths of commodity trade consisted of manufactures. At the same time, much of the trade was the supply of primary produce to the limited number of industrialized states in exchange for manufactures. In contrast, the rapid post-war growth in world commodity trade has been associated with a sharp increase in the share taken by manufactures – now about three-fifths of the total – and also in the volume of trade in manufactures conducted between the industrialized countries. This reflects the growing number of industrial countries (Rostow, 1978), greater product specialization

**Table 2.1**   Exports as a percentage of world GDP, main sectors

|  | 1970 | | 1980 | |
|  | World GDP (billion $) | Exports as % of GDP | World GDP (billion $) | Exports as % of GDP |
| --- | --- | --- | --- | --- |
| Agriculture | 181.9 | 31.8 | 602.8 | 45.3 |
| Mining and manufacturing | 726.7 | 29.9 | 2,769.5 | 55.0 |
| Services | 1,457.6 | 6.9 | 5,644.0 | 10.8 |
| TOTAL | 2,366.2 | 15.9 | 9,016.3 | 26.7 |

*Source*: Marshall *et al.*, 1988, p. 94

and also differentiation within product classes, leading to a sharp increase in intra-industry trade.

Throughout much of the nineteenth century and the first half of the twentieth, manufacturers in the industrialized nations regarded the domestic market as their own preserve. For consumer goods especially, but also over a wide range of investment goods, imports into the industrialized countries were regarded as the exception rather than the rule. It was widely held that to be successful as an exporter of manufactures, a country must have a strong and healthy domestic market. Much of the competition between the industrial nations lay in the markets of the less industrialized nations, including lands of recent European settlement, which paid for their imports by the export of primary produce. However, this competition was in practice limited in some degree by the policies of colonial powers and the operation of preferential trading arrangements; it was not until after the Second World War, with the attainment of independence by former colonies, that the greater part of these trade restrictions disappeared. Since 1945, the rapid growth of commodity trade has been most marked between the industrial nations. Note that the increasing relative importance of commodity trade is paralleled by a similar shift in respect of services, though the proportion traded is still much less than is the case for commodities. Nevertheless, in 1980 about one-quarter of total world GDP was exported (Table 2.1).

Although the evidence showing the post-war increase in the openness of national economies began to become available in the 1960s, it was not until a decade later that the fact began to be widely accepted. Two points are of special interest. Keynes published his *General Theory* in 1936 and his ideas were subsequently codified by Samuelson. At this time, it did not seem too unreasonable to treat a national economy *as if* it were closed, a convention that became implicit in much Keynesian thinking (Pratten, 1985, p. 65 ff.). On that assumption, an increase in demand that might be caused by government intervention would lead to higher output within the country itself. In practice, as Britain and other nations have discovered, a boost to consumer demand is apt to lead to a large inflow of imports and to the danger of a balance-of-payments crisis (p. 53).

In addition to cyclical surges in the import of manufactures, there has been a strong secular trend in the post-war period for imports to take a larger share of domestic markets. This phenomenon, known as  import penetration, has attracted the attention of numerous scholars, most notably Lord Kaldor (1971, 1981). If one assumes that the level of exports is given, an exogenously determined variable, then an increase in imports will indeed imply a lower level of activity in an economy than would otherwise occur. In which case, rising import penetration can be regarded as one of the symptoms and one of the causes of the economic malaise known as de-industrialization. Kaldor and others (e.g., Moore and Rhodes, 1982) have argued that, in order to raise Gross Domestic Product, some control of imports is essential. However, this policy prescription derives from an analysis which, concentrating on import patterns, ignores the fact that exports have been taking an increasing share of manufacturing output (Blackaby (ed.), 1981). Therefore, if a particular country is experiencing a level of import penetration which is sufficiently high to be creating problems, one has to ask why it is that exports from that country are not sufficiently buoyant. If this question is asked at the same time as we enquire about the reasons for rising imports, it is immediately evident that the phenomenon of crucial significance is the change in a nation's competitiveness – in both price and non-price terms. Furthermore, to examine just one sector of an economy – manufactures – without reference to the balance of trade in other sectors gives an incomplete, and often erroneous, view of the circumstances facing a country (Chisholm, 1985a; Rowthorn and Wells, 1987).

The debate about the importance of import penetration illustrates a fundamental matter. If one believes that manufacturing is primarily for the domestic market, with exports directed mainly to unindustrialized countries, then an increase in manufactured imports into an industrialized nation would indeed be a serious matter. However, export performance is as important as import performance, and both are related to competitiveness. This is the central fact which any theoretical analysis of national and regional economies must confront.

As national economies become more open, it is increasingly difficult to manage them as if they are in fact closed. This clearly has implications for what it is that governments can realistically hope to achieve, with respect to the management of national economies and also with respect to regional growth and decline. In addition, the distinction between a national and a regional economy for the purpose of economic analysis is reduced. This point was made implicitly by Brown (1972, pp. 72–6) when he compared the openness of the standard regions of the United Kingdom with the openness of selected national economies; the regions were more open than small countries such as Denmark and Norway, but the difference was essentially one of degree. At the national and international level, increasing interdependence poses two related problems of management: how to organize international affairs so as to maintain a reasonable degree of stability in global economic transactions; and how to manage national economies which must in fact be treated as open, whereas in the past the tendency was to regard them as being closed.

## Evolution of trade theory

As trade has gained in relative importance so have ideas concerning the theory of international and interregional trade evolved. While it would be wrong to present these changes as a linear evolution, there have been two unmistakable shifts. Trade was once conceived as a zero-sum game, but no longer; and whereas, in early formulations, comparative advantage was conceived in the static terms provided by natural resources and endowments of capital and labour, it is now generally accepted that comparative advantage is dynamic, dependent upon technological change and hence upon the qualities of the human actors. This latter idea, however, has only become generally accepted during the last three decades.

In the days of Mercantilist thought, prior to the publication of Adam Smith's *The Wealth of Nations* in 1776, it was widely held that trade was essentially a zero-sum game. Consequently, overseas possessions were to be exploited for the benefit of the metropolitan nation, and the primary objective of trade was to increase the national stock of specie. Adam Smith, building on the work of earlier theorists, challenged these ideas in a number of ways. Directly, he argued that it was both wrong and impossible to try to rein in the development of the American colonies (now the United States and Canada). He also set out the benefits to be had from the division of labour and hence the role of the market, and especially its size. The existence of scale economies in production implies that effective labour resources are not finite. Smith also articulated the idea that trade should be based on differences in production costs in the version we now know as the doctrine of absolute advantage. A re-allocation of production and trade on this basis would allow incomes to rise above their then current levels, but this increase would be limited by the resource base, which, he visualised, set an upper limit to the potential gains. Thus, although total resources were conceived to be finite, the world of Adam Smith was well within those limits. Trade was not thought of as a zero-sum game, and would therefore provide a mechanism for achieving the income levels that were realizable.

Forty-one years later, Ricardo formulated the more powerful concept of comparative advantage. This provides the rationale for trade between two or more countries when one country has an absolute advantage in producing both (all) of the commodities involved. If we suppose that consumers in all the countries wish to have access to all the commodities in the system, and if we assume that workers can freely shift from one occupation to another within a country but cannot emigrate, and if we assume zero transfer costs in international trade, it can be readily shown that countries should specialize in those products for which they have the greatest relative advantage. However, the Ricardian system, like that of Smith, is essentially static in character. Taking account of natural resources and labour supplies, the Ricardian doctrine holds that, once a reallocation of production and trade has occurred, the resulting pattern would be 'permanent'. Although Ricardo hinted that 'artificial' advantages might be important, he clearly considered that trade arose from 'natural' (i.e., permanent) advantages, such as location, climate and mineral resources, which affect the productivity of labour and capital. Implicitly, technology was regarded as fixed and unchanging.

Although Ricardo's ideas have formed the basis for all subsequent mainstream economic theorizing on trade, two exceptions to the idea of fixed technology were readily admissible. With the railway and steamship revolutions of the nineteenth century, transport costs fell dramatically and large areas were opened up to European settlement. With falling real transfer costs, the pattern of comparative advantage worldwide manifestly shifted. But this shift could be, and was, construed as changes in the ability to realize 'natural' comparative advantage as the 'distortions' created by transport costs were reduced. The second exception was the acceptance by some scholars, notably Ohlin (1933), of the existence of scale economies in production. Assuming that all countries have the same production function, the non-linear character of that function under scale economies implies that the relationship of production costs in two countries might vary over time for reasons unconnected with 'natural' advantages, especially in manufacturing. The case for protecting infant industries rests very largely on the existence of scale economies.

The exceptions, both of an apparently limited nature, did not seriously disturb the evolution of trade theory. The first major development from Ricardo's comparative advantage theorem is associated with Heckscher, Ohlin and Samuelson, whose contributions spanned the years 1914 to 1949 (Greenaway, 1983). Their theorem, known as the factor proportions theorem, asks the following question: Even if factor productivity is identical in various nations, can trade between them nevertheless arise? The answer is 'yes'. If labour and capital must be combined in the production process, if each commodity has its own production-function, if production-functions are identical in all countries, trade will arise if the endowment of labour and capital varies between nations. These differential endowments are taken as given.

Although the factor proportions theorem emphasizes labour and capital, rather than natural resources, it remains essentially static in concept. The supply of labour and of capital is taken as exogenously given, notwithstanding that capital is man-made. This basic assumption was not seriously challenged until the 1950s. The first step to this challenge occurred with the publication of two papers by Samuelson (1948, 1949), in which he derived the proposition that international trade should tend to equalize incomes. This conclusion was clearly at variance with the observed large differences, which were manifestly getting even greater; the contrast between the conclusions derived from theory and the evident realities suggested that the theory might be wrong. This paved the way for two seminal papers by Leontief (Johns, 1985). Leontief (1953 and 1956) sought, unsuccessfully, to substantiate the factor proportions trade theorem. The United States, being a capital-rich country, ought to export goods which embody much capital and import goods embodying large amounts of labour. The evidence did not support this expectation. Some scholars questioned the reliability of the evidence adduced by Leontief, but the more general reaction was to search for reasons which would explain the failure of the factor proportions theorem to yield the right predictions (Grimwade, 1989). Attention began to focus on the previously neglected fact that technical change is ongoing. As Nurske (1961, p. 34) put it: 'No useful purpose is served by continuing to discuss matters of trade and development on the classical assumption of a constant stock of productive factors.' Product and process innovations are a continuing feature of the world economy, and the quality of both fixed and of human capital is being improved. Therefore, although the factor proportions explanation for comparative advantage remains valid as *one* explanation for trade, it is at best partial and incomplete.

In the same year (1961) that Nurske wrote, Linder and Posner offered explanations for the paradox identified by Leontief. Linder argued that trade theory hitherto had concentrated exclusively on the cost of production, which is a matter of supply, and had neglected the role of demand. He pointed out that consumer preferences do vary from one country to another, and that there is a systematic variation in these preferences with the level of personal income. He argued that countries produce primarily for the domestic market and only engage in exports when that market is saturated. Consequently, each country, he argued, produces its own special version of individual commodities, for which there will be a market niche in countries at an equivalent level of development. On this basis, industrial nations will trade with each other.

While Linder's thesis is somewhat forced, his emphasis on demand is important, and can be generalized to the idea of product differentiation. This in turn is consistent with Posner's alternative formulation of the missing ingredient, as the idea of the technology gap. It is quite evident that the quality of technology which is actually applied does vary from country to country, such variation being only loosely correlated with the stock of capital as conventionally defined. Consequently, production-functions are variable across space, not uniform. One aspect of the technology gap is the idea of the product cycle, generally associated with Vernon (1966; but see also Hufbauer, 1966). After the initial commercial development of a product, manufacturing technology usually stabilizes. This provides scope for nations other than the one of first adoption to enter production, quite possibly undercutting and then eliminating the pioneer.

Given the fact of technological change, the rate at which capital investment occurs takes on significance. In advanced industrial nations, it is reasonable to assume that new capital goods embody the current state-of-the-art technology and that, consequently, they will give higher productivity or better quality, or some combination thereof, than can be obtained from capital of an older vintage. The average age of the capital stock will therefore have a bearing on the efficiency of production. If investment in competing nations (regions) occurs at the same rate, no competitive advantage is available to anyone. If, on the other hand, one nation (region) has a higher rate of replacement and/or of net addition to capacity, that nation (region) will achieve an advantage in unit costs relative to its competitors. This effect, known as Verdoorn's Law, means that a nation which does not have a comparative advantage, or is threatened with the loss thereof, may be able to create an advantage by a high rate of investment and the acquisition of current best practice technology. As with the scale effect, Verdoorn's proposition applies at the level of the individual plant (firm) and also in the more general spatial sense.

With the acceptance of technological change as an important source of comparative advantage, two things fall into place. Transport innovations and scale economies cease to be exceptions to the general assumption of static technology and become but two aspects of the whole dynamic process. The implication is that production-functions will vary across space and can no longer be assumed to be homogeneous. In the second place, product differentiation provides the basis for intra-industry trade of a more general kind than envisaged by Linder. The rapid growth of product differentiation and intra-industry trade provides much of the basis for the post-war expansion in trade in manufactured goods between the industrialized nations (Bolassa, 1966; Williamson, 1983).

Thus, we can now recognize four primary sources of comparative advantage: natural resources; endowment of capital and labour; technological leads and lags; and product differentiation. All four explanations have in fact been valid virtually since trade began. However, the changing perception concerning the reasons for trade probably reflects the evolution of the world's commercial structure, in particular the growing importance of man-made as compared with natural advantages as an explanation for trade (see, for example, Tyson and Zysman, 1983). For our purpose, however, it is important to realize that this changing perception has occurred only during the last three decades or thereabouts.

To what extent the man-made element of trading advantage will continue to gain in relative importance is a matter for speculation. In a thoughtful review of events since his 1966 paper, Vernon (1979) has drawn attention to three trends which seem to be blurring the operation of the product cycle (which is an important part of the technology gap basis for trade). Vernon points to multinational enterprises (MNEs), each with a presence in an increasing number of countries; the worldwide availability of knowledge to many innovating firms has thereby improved considerably. At the same time, the per caput income lead enjoyed by the United States over Europe and Japan has been dramatically closed, with the corollary that the United States is no longer the pre-eminent source of industrial innovation. The third notable change has been the move to product differentiation and market segmentation in some industries (for example, automobiles) and the declining relative importance of industries with standardized products that could look to worldwide markets (for example, petroleum and steel). Although these three trends have blurred the operation of the product cycle, and possibly reduced its relative significance, they do not vitiate the principle, that the 'roles of innovation, scale, ignorance and uncertainty' (Vernon, 1966, p. 191) materially modify the location and trade patterns predicted by comparative advantage analysis based on natural resources and the endowment of capital and labour.

The net effect of these changes, and the increasing openness of national economies, has been summed up as follows for the United States, which is by far the largest single productive and trading system in the world:

In a simple quantitative measure, the shares of imports and exports in U.S. manufacturing value-added both more than doubled from 1960 to 1980. But the change was more than a quantitative one – it amounted to a qualitative change in the importance of international considerations to the U.S. economy. In 1960 the typical U.S. manufacturing firm was basically oriented toward selling to U.S. consumers and competing with U.S. rivals. If it exported, this was usually a secondary activity; if it faced foreign competition, this was usually a minor irritant. By contrast, in the 1980s international considerations have become a key factor. Many, perhaps most, firms either rely heavily on export sales or face important foreign competition in the U.S. market.

(Krugman, 1986, pp. 5–6)

With differences in timing, much the same can be said of all the major industrial nations. Hence:

... comparative advantage and competitiveness are dynamic rather than static ... Capital and human capital are created resources, and a nation's competitive position is thus determined in part by its relative success in creating resources.

(Arndt and Bouton, 1987, pp. 34–5)

## The sources of economic growth

Just as ideas concerning the sources of comparative advantage have been in flux, so also have notions with respect to the springs of economic growth. The accumulating evidence on the sources of growth has resulted in a shift of thinking which roughly coincides in time with the post-war evolution of trade theory, shares the same intellectual inspiration and leads to similar conclusions.

Classical economists believed that the wealth of nations was largely determined by the abundance and quality of the natural resources – the supply of minerals and fuels and, especially, the fertility of the soils. In the economists' shorthand, the pre-eminent factor was 'land'. This view was superseded in the nineteenth century by an emphasis upon the expansion of the labour force and the accumulation of capital as the main determinants of the rate of growth of national economies. Capital, in this context, means the productive capacity represented by factories and machinery, infrastructure investment and the like. This tradition carried into the post-war period, most notably in Rostow's 1960 essay on the stages of growth. He hypothesized that a necessary condition for self-sustained growth is the achievement of an adequate level of investment, expressed as a proportion of GNP. Notwithstanding the caution which he expressed, his ideas were seized upon and converted into the widely believed proposition that a suitable percentage of GNP devoted to investment would ensure that sustained growth would occur.

However, soon after Rostow's influential essay appeared, Denison (1967) and Kuznets (1966, 1971) published major studies in which they sought to identify the sources of economic growth, building on earlier work dating from the 1950s (Thomas, 1985, p. 15). Using the conventional categories of land, labour and capital, the evidence pointed very clearly to the following conclusions for the more developed countries. Land as a factor of production is difficult to measure and much of its contribution is made through the productivity of capital and labour. Therefore, in terms of measurable contributions, the spotlight fell clearly on labour and capital as the potential sources of growth in national output. With long-run data spanning somewhat over a century in several cases, expectations were broadly confirmed for the nineteenth and early twentieth centuries, but not for more recent times. One summary of the evidence expresses the position thus:

> Whereas in the earlier periods, over half of the growth in total output could be attributed to increases in the inputs of labour and capital, in more recent times much the greatest increase has arisen from higher productivity.
>
> (Chisholm, 1982, p. 44)

Kuznets (1971, pp. 306–7) estimated that productivity increases accounted for 80 per cent of the rise in per caput product in many countries.

Productivity increases depend not just on the quantity of capital available for each worker but also on the quality of the capital, the manner in which work is organized and the quality of the labour force (health, education, skills and attitude). To the extent that capital goods are themselves improving over time, this will be because they embody practical experience and also scientific or technological advances. Some of these improvements can be obtained for little cost, and may actually reduce the outlay needed to achieve a given output.

The work of Denison and Kuznets amounted to a powerful attack on the then conventional wisdom regarding the sources of economic growth, shifting the spotlight from the quantity of labour and capital to changes in productivity attributable to technological change and innovation. Other scholars have attempted similar studies, to see whether the Denison/Kuznets findings can be confirmed or rejected. Possibly the most careful of these is a study by Jorgenson, Gollop and Fraumeni (1987) of the United States' economy from 1948 to 1979. Growth in value added, averaged over the entire period, was as follows:

|                                  | Annual % growth |
|----------------------------------|-----------------|
| Value added                      | 3.42            |
| Contribution of capital input    | 1.56            |
| Contribution of labour input     | 1.05            |
| Rate of productivity growth      | 0.81            |

These figures suggest that most of the growth in value added is attributable to expansion in the amount of labour and capital and appear to contradict the Denison/Kuznets findings. However, these contributions include the estimated improvement in the quality of both of these factors of production. Thirty-one per cent of the capital component was attributed to quality improvements; for labour, the equivalent proportion was 28 per cent. If the figures are now recalculated, the overall 'quantity' component for labour and capital becomes 1.84, and that for quality plus productivity 1.58, summing to the annual growth of 3.42 per cent. These authors, therefore, give productivity and quality a lower significance than other authors, but nevertheless a role which accounts for about 46 per cent of overall growth.

Hulten and Schwab (1984) report a rather similar finding, also for the United States but for the slightly shorter period of 1951 to 1978. Value added in manufacturing increased at 3.31 per cent per annum over that period, with 1.70 per cent per annum arising from productivity increases, i.e., 51 per cent of the total. Furthermore, they also confirm the finding by Jorgenson and his co-authors that increases in the quantity of labour account for only about one-fifth of the rise in output, implying that some 80 per cent is due to elements of the economic system which are in large measure under the control of governments and corporate decision-makers, i.e., the quality of labour and the quantity and quality of capital.

The situation becomes more complex at the regional level, since interregional migration materially influences the rate at which the regional labour force expands and the volume of matching investment which is necessary. As Table 2.2 shows, the Sunbelt regions of the United States have been expanding much more rapidly than the Snowbelt regions, but most of this differential arises from the increase in the labour force, which is largely occasioned by inward migration. Output per worker has actually been rising more quickly in the Snowbelt than in the Sunbelt, indicating the long-term potential which the Snowbelt has to maintain and even enhance its relative personal income and hence attractiveness as an area in which to live and work. Expressed in a somewhat different way, it is clear that regional variations in the growth of productivity, as distinct from the quantity of the factors of production, has a material impact on the geography of growth and prosperity.

These findings confirm the evidence discussed in the previous section concerning the sources of comparative advantage, and reinforce the view expressed by Becker (1964) and Schultz (1961, 1981) that human capital is the single most important resource. On this view, it is investment in the health of workers and in their education and aptitudes which provides the mainspring of development. Consequently, national and regional fortunes are attributable in large measure to the qualities and abilities of the people, individually and collectively, rather than to some 'permanent' conditions either of location and resource, or of the quantity of capital and labour inherited from a previous generation.

## The changing organization of industrial production

The sharp growth of international competition in manufactures during the post-war period creates conditions in which even big, and seemingly dominant, firms cannot regulate and control their markets in the manner that had been customary before the Second World War. For this reason, Piore and Sabel (1984) argue that manufacturers are being forced to abandon the inflexible, mass-production (Fordist) mode of operation and opt for more flexible systems, so that responses to market signals can be quick and precise. At the

**Table 2.2** Sources of output growth and labour productivity growth in US regions 1951–78 (% growth per annum)

| Region | Output growth | Growth of output due to: Capital-stock growth | Growth of output due to: Labour-force growth | Productivity |
|---|---|---|---|---|
| Middle Atlantic | 1.78 | 0.44 | −0.35 | 1.70 |
| New England | 2.24 | 0.48 | −0.06 | 1.82 |
| East North Central | 2.62 | 0.68 | 0.14 | 1.79 |
| West North Central | 4.19 | 1.20 | 0.97 | 2.02 |
| *Snowbelt regions* | 2.45 | 0.62 | 0.03 | 1.80 |
| South Atlantic | 4.49 | 1.42 | 1.39 | 1.69 |
| Pacific | 4.76 | 1.26 | 1.84 | 1.67 |
| East South Central | 5.09 | 1.90 | 1.58 | 1.61 |
| West South Central | 5.59 | 2.40 | 1.77 | 1.42 |
| Mountain | 5.87 | 1.90 | 2.58 | 1.39 |
| *Sunbelt regions* | 4.94 | 1.63 | 1.69 | 1.61 |
| TOTAL | 3.31 | 0.96 | 0.65 | 1.70 |

*Source*: Hulten and Schwab, 1984, p. 157, modified by Armstrong and Taylor, 1985, p. 60

same time, the pace of change has also quickened with the advent of so-called 'high technology' manufacturing – electronics, computer-based products and biotechnology, for example. With new products coming to market quickly, and with new ways of doing old things, the need for flexibility has been sharply increased. Two developments have made it possible to respond to this pressure: technological change itself and the rethinking of organization structures and work practices.

Numerically controlled machines and robots permit production lines to change from one product or model to another with minimal disruption of the flow of work. Long runs of identical goods are no longer as necessary as was formerly the case. At the same time, modern information systems allow production and stocks to be geared much more accurately to customers' needs, reducing or eliminating the wholesale and warehouse function in the distribution system. For some products, a customer can place his order through a local retailer direct to the manufacturer, for delivery to the retailer within a matter of days. The same technological advances make it possible to organize the input of components in a much more responsive manner than hitherto. The full advantage of these technological changes can only be obtained, however, if the workforce itself can operate in a flexible way, which implies reducing or eliminating restrictive work practices, and if the whole pattern of sourcing is reconsidered. Some firms, most notably in the vehicle industry, have adopted just-in-time systems for the supply of components from subcontractors. At the same time, thought must be given to the proportion of components which is

**Table 2.2** (continued)

| | Growth of output per worker | Growth of output per worker due to: | |
| --- | --- | --- | --- |
| | | Growth of capital/labour ratio | Productivity |
| Middle Atlantic | 2.25 | 0.56 | 1.70 |
| New England | 2.26 | 0.45 | 1.82 |
| East North Central | 2.34 | 0.55 | 1.79 |
| West North Central | 2.70 | 0.68 | 2.02 |
| *Snowbelt regions* | 2.36 | 0.56 | 1.80 |
| South Atlantic | 2.31 | 0.62 | 1.69 |
| Pacific | 2.15 | 0.48 | 1.67 |
| East South Central | 2.45 | 0.83 | 1.61 |
| West South Central | 2.45 | 1.03 | 1.42 |
| Mountain | 1.91 | 0.52 | 1.39 |
| *Sunbelt regions* | 2.31 | 0.70 | 1.61 |
| TOTAL | 2.33 | 0.63 | 1.70 |

bought from subcontractors as compared with the proportion produced internally.

There has been a very sizeable literature which seeks to explore the significance of these and related changes for the geography of production. Four strands of thought are particularly noteworthy. First is the view that the rise of new industries implies the development of new industrial areas and the decline of old ones. The changing circumstances of production are said to be readily mapped onto the rise and fall of regional fortunes (e.g., Hall, 1985; Markusen, Hall and Glasmeier, 1986; Scott, 1988a, 1988b). Second, and more nearly based on the product cycle concept, is the belief that production is being rapidly reorganized internationally, especially through the operations of multinational enterprises, so that routine operations are located where labour costs are low. The so-called New International Division of Labour is thought to be a powerful factor causing the de-industrialization of advanced economies (Frobel, Heinrichs and Kreye, 1980). At the intra-national level, Massey (1984) has drawn attention to shifts in the location of production reflecting the skill requirements of processes and the labour and other characteristics of regions. Finally, it has been argued that flexible production systems in which just-in-time sourcing is important imply the advantages of proximity, and hence a reassertion of the economies to be gained from agglomeration (Piore and Sabel, 1984; Scott, 1988a, 1988b), to which must be added the advantages proximity to markets gives in responding to market place trends (Schoenberger, 1986, 1987).

Gertler (1988) has raised pertinent questions concerning the nature and the scale of the changes which are currently affecting manufacturing industry, suggesting that the shift from 'Fordist' to 'flexible' production has been exaggerated. Rigid mass-production systems have in fact coexisted with more flexible production, and on some estimates only about 20 per cent of manufacturing output was ever organized on truly Fordist lines – the vehicle industry itself being the type case. The single best review of the rapidly growing literature is by Milne (1989), whose own work on selected industries in Britain shows that adoption of flexible shop-floor production systems is associated with complex patterns of adaptation. Consequently, simple statements and generalizations are not possible. For example, some firms achieve flexibility by externalizing the supply of components, while others build new premises, with modern equipment, to bring the source of parts back under their own direct control. Some production which had been moved abroad has now returned to the United Kingdom, though not always the same firms.

While there is no doubt that important changes have occurred and continue to occur, a simple dichotomy between Fordist and post-Fordist production is misleading and inappropriate. At the present time, though, the pressure on firms is toward more flexible and responsive manufacturing systems, in a way that was less evident in the 1950s and 1960s. This shift of emphasis is a response to the combination of: more intense competition in domestic and overseas markets; rapid product and process innovation; and the assertion by customers of their desire to have the goods of their choice at the time of their choosing. Viewed from the regional perspective, this all adds up to the following proposition. Whereas formerly the comparative advantage of a region depended very largely on the availability of materials and fuels, upon labour skilled in particular operations and upon the cost advantages accruing from scale economies, the prime (though not the only) requisite today is the ability to provide the right environment for innovation and for flexible production.

## Structural changes

Some half-century ago, Clark (1940) and Fisher (1935) drew attention to the shifting structure of employment as economic growth proceeds. Employment in agriculture and the other primary industries falls, both absolutely and relatively, while the number engaged in manufacturing and services tends to rise. As development proceeds, however, the share of employment taken by manufacturing also falls, so that in the long run it is the service sector which increases in significance. These empirically observed trends can be explained in terms of the income elasticities of demand. As incomes rise, the proportion spent on 'necessities' declines, with a matching increase in outlays on goods and services considered desirable but not necessarily essential. The reality of the shifts noted by Clark and Fisher has been confirmed by many subsequent studies (e.g., Hall, 1987a – see Table 2.3).

As advanced economies become more dependent upon services, it might be thought that the overall significance of international and interregional competition would decline. In the Clark/Fisher formulation, services were visualized as ministering primarily to final consumption, in which case provision

**Table 2.3** Industrial employment, selected countries (percentage distribution, 1963–83)

|                  | United States | | | United Kingdom | | |
|------------------|------|------|------|------|------|------|
|                  | 1963 | 1973 | 1983 | 1963 | 1973 | 1983 |
| Agriculture, etc.| 7.1  | 4.2  | 3.5  | 4.4  | 2.9  | 2.7  |
| Industry         | 35.1 | 33.2 | 28.0 | 46.4 | 42.4 | 33.6 |
| Services         | 57.8 | 62.6 | 68.5 | 49.2 | 54.6 | 63.7 |
| TOTAL            | 100.0| 100.0| 100.0| 100.0| 99.9 | 100.0|
|                  | West Germany | | | Japan | | |
|                  | 1963 | 1973 | 1983 | 1963 | 1973 | 1983 |
| Agriculture, etc.| 11.9 | 7.3  | 5.6  | 26.0 | 13.4 | 9.3  |
| Industry         | 48.7 | 47.5 | 42.0 | 31.7 | 37.2 | 34.8 |
| Services         | 39.3 | 45.2 | 52.4 | 42.3 | 49.4 | 56.0 |
| TOTAL            | 99.9 | 100.0| 100.0| 100.0| 100.0| 100.1|

*Source*: Hall, 1987a, p. 96

must be local and accessible to households. In addition, public administration (central and local government and public agencies) is, in important ways, geared to the distribution of people as consumers. For many years, services have been widely regarded as 'unproductive' and dependent for their existence on the 'productive' primary and secondary industries. Such a view ignores the potential which services have for enhancing the quality of life, the overseas package holiday industry being a case in point. Equally important, though, is the fact that some services minister to the needs of producers, not final consumers. Transport services and insurance, for example, have long been recognized as intermediate inputs in the manufacturing process. Less recognition has been accorded to the increasing range and importance of producer services in all the advanced economies.

These developments have been admirably documented in some recent publications (Daniels, 1985; Marshall *et al.*, 1988; and Wood, 1986). Marshall and collaborators note, for example, that the 1981 census for Great Britain recorded 8.6 per cent of the employed population working in producer services, with a further 12.9 per cent in industries which cater partly for producers and partly for final consumers. Although some consumer services are traded internationally, the majority are oriented to local consumers and are largely insulated from direct international competition. In contrast, producer services are traded between regions and internationally. Furthermore, since producer services provide inputs to the production industries, and since the latter are caught up in worldwide competitive struggles, producer services are fully engaged in the processes and pressures which arise from the increasing openness of national economies.

Table 2.1 shows the composition of world GDP and the share that enters international trade. Almost two-thirds of world GDP is accounted for by services of all kinds – consumer services and producer services. Of the service output, almost 11 per cent was traded internationally in 1980. Although the significance of international trade has been increasing more rapidly for manufactures than for services, the table shows that the proportion is rising quite substantially for the service sector.

There is little doubt that the world economy is on the threshold of a considerable expansion in the international trading of services. Thus it is that the structural shift of many advanced economies, away from manufacturing and to services, should not be taken to mean that the competitive pressures identified in preceding sections will abate. Indeed, the evidence of recent years is that these competitive pressures are becoming greater in the service sector, so that, whatever the economic basis for a region's economy, the need to be competitive and adaptable is becoming greater.

## Multinational companies

Previous sections have documented the growth of international trade as an increasing proportion of output, for both goods and services, with the implication that the environment for firms is becoming ever more competitive. That implication might be false, given that in recent decades the role of the multinational corporation has expanded dramatically. As Taylor and Thrift (1982, p. 1) put the matter:

> The 1970s can now be seen as a period in which a major shift took place in the structure of the world economy which can be interpreted as either the start of, or a period of transition to, a new phase of internationalization of production.

This 'internationalization' is manifest in the rise of what they call the 'global corporation', more usually known as multinationals or MNEs. By merger and acquisition, and by the establishment of new offshore plants, the post-war era

**Table 2.4**  Intra-firm trade as a percentage of foreign trade

| Country | Year | Imports | Exports |
|---------|------|---------|---------|
| Canada | 1964 | 32 | - |
|  | 1969 | 40 | - |
|  | 1971 | - | 59 |
| Sweden | 1975 | - | 29 |
| United Kingdom | 1973 | - | 30 |
| United States | 1970 | - | 50 |
|  | 1974 | 46 | - |

*Source*: Linge and Hamilton, 1981, p. 65

has indeed witnessed the growing importance of firms which have production facilities in ten, twenty, even forty countries or more. The share of output and of international trade which multinational companies control has been steadily rising.

*The Economist* (4 April 1982) estimated that approximately one-quarter of world commodity trade consisted of transactions internal to individual multinational companies, with the implication that their total involvement must exceed that proportion substantially. Table 2.4 records some estimates for individual countries in the 1960s and 1970s which are clearly consistent with the overall estimate for intra-firm trade in the early 1980s. If one takes the share of national output, Dunning (1981, p. 4) lists six advanced countries in which one-third of manufacturing and/or one-half of the primary sector is accounted for by multinational companies. These are:

| | |
|---|---|
| Australia | Ireland |
| Belgium | Norway |
| Canada | Sweden |

Finally, the bigger multinational companies are very large indeed. Taking the GNP of countries, and consolidated sales of companies, for the year 1976, of the first one hundred in the world, forty were multinational companies, many of which had sales in excess of the GNP of nations (Linge and Hamilton, 1981, pp. 50–1). On this basis, the economic power of some multinationals is truly colossal. Overall, multinational companies account for over 40 per cent of world output, and about one-quarter of the work force in the industrialized nations (Strange, 1988, p. 74).

Although some of the larger multinational companies operate in the petroleum industry and other extractive industries, it is the expansion of multinational companies in manufacturing which has attracted most attention in recent years. Thus, for the majority of advanced nations, the manufacturing sector, and trade in manufactures, is strongly influenced by the activities of multinationals. This fact is implicit in the data in Table 2.4, especially the exports of countries such as the United Kingdom and the United States.

The first multinationals were based in the more advanced countries, and the greater number of such companies is still based there. The more advanced nations provide the main sources of offshore investment. As a result, multinational companies have been closely involved with the rapid growth of trade in manufactures between the industrialized nations which has been a striking feature of the post-war period. In addition, these companies have played a significant role in the industrialization of some less developed countries, Singapore being a notable example. Finally, multinational companies are now being established in Brazil, India, Korea and elsewhere, even though the size and number of these firms are both small in comparison with those based in the United Kingdom, United States, West Germany, etc. Multinationals have become a truly global phenomenon (see, for example, Lall, 1981, 1983; Taylor and Thrift (eds), 1986).

It might be argued that the rise of the multinationals constitutes a severe erosion of real competition, which more than offsets the increase in the competitiveness of the environment in which firms find themselves on account of the increased proportion of output which is exported. However, the recession which many countries experienced between 1979 and 1983 created fiercely competitive conditions in which multinational companies were driven to rationalize and reorganize, demonstrating that they could not control and manipulate the situation but perforce had to adapt to survive. Survival strategies included the closure of plants and the relocation of production, a matter, very often, of shifts between countries. Thus, the alternative view of multinationals is that they could survive and prosper by virtue of their ability to switch investment with relative ease to the most advantageous locations. At the extreme, it has been suggested that multinationals possess capital which is hyper-mobile, i.e., mobile at zero or negligible cost. While this view is open to challenge, especially in respect of fixed investment, it does imply that multinational companies, far from being exempt from competitive pressures, are merely better able to adjust than is the case with companies that operate only within a nation or even just a region.

In the present context, two features of the rise of the multinationals are of particular relevance. First, the image of firms as atomistic units, as portrayed in the economists' perfectly competitive world, is clearly not true of a large segment of economic activity. As a consequence, the geography of production and trade is shaped by the decisions which are taken at head office, perhaps thousands of miles from the focus of the activity itself. Many observers have interpreted this to imply a loss of 'control', at local and even national level, over the disposition of resources. However, if in fact multinational firms operate in a competitive environment, it is not self-evident that the outcome of decisions taken elsewhere will necessarily differ from what would have been the case were firms to be locally based. The second point is connected. To the extent that governments seek to intervene in the allocation of resources, they are faced with some firms which are genuinely able to compare situations in more than one country (region) and to bargain for extra subsidies or other concessions. The appearance, therefore, is of the erosion of effective national control over investment and output decisions. However, the real erosion of such control lies with the increase in economic interdependence of countries, reflected in the proportion of output which is traded, of which the growth of multinationals is itself a symptom. For however local the decision process, no firm can escape the implications of developments and decisions elsewhere within the global economy.

Although multinational companies have attracted a good deal of attention, being an important feature of modern economies, small firms have not been eclipsed entirely. Indeed, as will be discussed in Chapter 8, there has in recent years been a marked increase in their number and importance. The trend toward an ever more dominant role for multiplant companies seems to be in abeyance, at least for the present.

## Conclusion

We have reviewed several strands of evidence and bodies of associated thought, all of which point in the same general direction. Taking a long historical view, there is no doubt that the environment in which firms operate in the industrial nations has become a good deal more competitive since the Second World War than it had ever been previously. The key fact is the marked increase in the significance of international trade relative to total output, one symptom of which is rising import penetration. The associated increase in trade in manufactures between the industrial nations cannot be explained by conventional factor proportions theorizing. As a result, trade theory has been modified to take account of technological leads and lags, and product differentiation. In other words, man-made advantages and mutable advantages have been fully accepted into doctrine explaining the patterns of comparative advantage. At the same time, work aimed at elucidating the sources of growth, as distinct from the spatial allocation of output and trade, has shown that a major component is the quality of capital and labour plus the 'costless' benefits of changes in socio-economic organization. The 'permanent' advantages of land resources and stock of both capital and labour are now seen to be less significant than previously thought. Associated with these changes are the present emphasis on flexible production systems on the one hand, and the rise of producer services on the other.

All of these considerations are relevant for the way in which we view regions and their economies. Perhaps the main points of relevance are the following. First, as national economies become more open, so they become less readily distinguishable from regional economies; regions and nations become more clearly aligned along a continuum, rather than being discretely distinct. Second, the fortunes of a region must depend in part on local circumstances and in part on conditions which are external. To the extent that it is local, or endogenous, factors which are responsible for success or failure, they lie to an increasing extent with the people of the region – the entrepreneurs, the workers and the administrators – rather than with the natural resources or, even, with the differential costs of transport associated with location.

Much of the traditional theory concerned with regional growth was developed with little or no reference to these wider changes, while some of the more recent attempts to develop appropriate theories of regional development have also ignored their significance. Equally important is the fact that much theorizing on regional growth processes has been based on bodies of economic doctrine which developed before the Second World War. To the extent that economic doctrine is a creature of its time, appropriate, one hopes, for that time, and to the extent that ideas have then been carried forward to an era which, in material senses, is different, it is possible that some irrelevant or erroneous doctrine may be accepted as true. Perhaps more important, though, is the problem that any particular way of looking at things is likely to be of limited utility, valid only for specific circumstances. Consequently, the next step in our enquiry is to examine the major strands of economic debate, to identify the main schools of thought. These all have implications for the way in which one approaches regional problems; the evolution of economic thought, therefore, has a direct bearing on the conceptualization of regional matters.

# CHAPTER 3

# *Some economic debate*

Like the nocturnal drunkard who searches for his lost keys under a streetlight because that is where they are easiest to see, so national economies looked for solutions to their problems in the areas illuminated by their past practices.

(Piore and Sabel, 1984, p. 222)

Ideas concerning regional growth processes are largely derived from more general economic doctrine. However, that is not a single, accepted body of thought but divides into a number of schools, each of which is distinctive by virtue of that which it emphasizes or ignores. Each school of general economic doctrine has implications for the way regional growth processes are visualized, either because general theory can be mapped directly into the spatial domain, or because of the indirect implications. Therefore, it is necessary to give some consideration to the nature of the schools of economic doctrine and to indicate, in a preliminary way, the manner in which the regional dimension may be treated. In so doing, it is important to note the time when a doctrine was promulgated and had its main impact, since this helps to elucidate the way in which the analysis of regional economies has evolved.

Five schools of economic thought may be identified which are of special relevance to us, as in Table 3.1. Keynesian ideas of economic management dominated the 1950s and 1960s, and had a profound and pervasive impact on regional economic thought. During this period, neo-classical ideas were largely in eclipse, though often used as a basis for comparison with Keynesian concepts. During these two decades, location models derived from the principles of monopolistic competition achieved prominence, but suffered from the problems inherent in converting micro-economic analysis into aggregate, and growth-oriented, terms. Since the early 1970s, the ascendancy of Keynesian ideas has been challenged on two fronts. The more successful challenge has been mounted by economists persuaded of the virtues of monetarism and the importance of the supply-side of the economy. Although there has been a remarkable growth in supply-side *practice* at the regional level, nobody has as yet fully formulated ideas of regional growth in supply-side terms. The other challenge has been mounted by radical-minded scholars persuaded of the merits of the writings of Marx, Lenin and others in that tradition.

If there is a single proposition which can be made about the discussions in the literature, it is the following. The Keynesian consensus emphasized the role of

**Table 3.1**   Schools of economic thought and regional analysis

| Schools of thought | Important authors | Regional analysis Period of dominance or significance | Important authors |
|---|---|---|---|
| Neo-classical | Marshall Walras Knight | Dominant late nineteenth century to 1920s | Regional problems generally not perceived |
| | | Continuing influence | Borts & Stein (1964) Richardson (1973) |
| Monopolistic competition | Chamberlain Robinson Ohlin | Significant 1950s and 1960s | Christaller (1933) Greenhut (1956) Hotelling (1929) Isard (1956) Lösch (1954) |
| Keynesian | Keynes Samuelson | Dominant 1950s and 1960s | Brown (1972) Kaldor (1970) |
| Neo-Marxist/ radical | Marx Lenin Harvey | Significant 1970s and 1980s | Frank (1964) Harvey (1982) |
| Monetarist/ supply-side | Friedman | Significant 1970s and 1980s but not really incorporated into theory | |

demand management to the virtual exclusion of supply-side issues. In the regional context, regional economies were viewed as dominated by the exogenous demand for their products or by the exogenous injection of demand through government intervention. So far as supply-side issues were considered, they were treated in neo-classical terms – population as an exogenous factor and the mobility of (homogeneous) capital and labour in response to differing rewards, but tending to equality in equilibrium. Supply-side issues of a more disaggregated nature were largely ignored, as also the question of technological change and the role of the entrepreneur in fostering economic growth. Since the early 1970s, and more especially during the 1980s, that neglect has been partially rectified, as a response to the monetarist/supply-side challenge and to the evidence of what is actually hapening 'on the ground'.

Although economic ideas have not evolved in a simple linear progression, it is useful in this chapter to treat the five schools in the quasi-historical way suggested in Table 3.1. Such a treatment should not be taken to imply that the 'later' schools are superior to the 'earlier' ones. Nor should it be taken to suggest that in each case the more recent school replaced the preceding one. It seems more relevant to think of ideas being promulgated which, at least in part, are either incorporated into later thinking, or coexist.

In presenting an outline of economic doctrines, every attempt will be made to treat each corpus of ideas in a fair and 'detached' manner. However, it is impossible to conceal one's predilections, and unwise to pretend that they do not exist. At the most general level, each major school of economic thought is based on assumptions, taken as premises, which, in the generality of cases, can only be described as heroic. On the other hand, if the doctrines are described, it is possible to recognize the sources of ideas relevant for regional analysis and the way in which these ideas have evolved.

## Neo-classical theory

Writing of the period 1870 to 1914, Schumpeter (1954, p. 952) remarked: 'there existed by about 1900, though not a unified science of economics, yet an engine of theoretical analysis whose basic features were the same everywhere.' Schumpeter himself objected to the description of this corpus of theory as 'neo-classical'. Nevertheless, this term has become widely used to represent the work of Clark, Marshall, Walras, Wicksell, Wicksteed and others (Harcourt, 1972). This neo-classical school differed from the preceding classical one both in the focus of interest and in the analytical tools employed. Regarding the former point, the contrast has been succinctly put in the following terms by Pasinetti (1977, p. v): 'The theory of production was at the very foundations of Classical economic theory but was later pushed into a secondary and subordinate position by the development of the marginal theories of consumers' behavior and of exchange in competitive markets.' Central to the neo-classical endeavour was the elaboration of the concept of utility and of the conditions for equilibrium. The Walrasian system of simultaneous equations provided the culmination to the neo-classical endeavour, made possible by the use of tools relatively unfamiliar to earlier generations of economists – the tools of mathematics.

Neo-classical theory was essentially micro-economic – concerned with individuals and firms, and their behaviour. It was clearly recognized that behaviour (as buyers, sellers, producers or consumers) is conditioned by the actions of all the other micro-economic actors as they respond to market signals, i.e., prices. However, it was assumed that the 'normal' condition of markets approximated those specified in the 'perfect competition' economic model, i.e., that it was both usual and desirable for economies to be either in equilibrium or tending toward an equilibrium state, and that this would be an equilibrium in which labour would be fully used, being a condition of full employment. Consequently, the central problems were to specify the circumstances needed to attain equilibrium, to consider the distributional and equity implications, and how it is that individuals and firms make decisions on the allocation of resources. With price signals being the key decision stimulus, the theory of price was of crucial importance; the determination of price would determine output (Deane, 1978).

The assumption that an economy tends to equilibrium at the full use of resources carries the implication that total output will be determined by the supply of the factors of production. Because the work force in a closed economy will change in size only with a lag of about fifteen years – depending, as it would do, on changes in net reproduction – it is customary to think of aggregate production as being determined by the number of people of working age. In this sense, the neo-classical model is said to be supply-determined. However, this attribution is valid only in the short term, since reproduction can and does change in the long run. Consequently, to describe the neo-classical model as supply-determined begs the question of what it is that controls the supply of people, i.e., it leaves out of the account any theory of population change. Therefore, it would be more accurate to describe the neo-classical model as one in which total output is determined by the supply of labour, which is taken as exogenously given.

To the extent that neo-classical scholars considered the sources of economic growth, they assumed that output would grow in proportion to the increase in population and to technological change, both treated as exogenous variables. If, for example, population increased by 1 per cent per annum and productivity rose at the same pace, total output would grow by 2 per cent each year. Clearly, this is a very simple view of the growth process.

Neo-classical economics was constructed on a triad of production factors – land, labour and capital – and thereby maintained a long tradition. Furthermore, with only limited exceptions (see Schumpeter, 1954, pp. 899 ff), scholars conceived these factors as being homogeneous, notwithstanding the fact that all three are heterogeneous in practice. However, the assumption of homogeneity is necessary if there are to be continuous supply and demand schedules with tractable properties, and if there is also to be a continuous substitution possibility between factors (especially labour and capital). Despite the critics, this belief in the homogeneity of the factors of production remains deeply embedded in economic thinking.

The micro-economic framework stresses the idea of supply and demand schedules and their intersection to give a determinate price, the idea of production-functions facing firms, and the idea of utility functions for individuals. For there to be a determinate equilibrium situation, it is assumed that producers and consumers can, and actually do, move along the relevant schedules, and that adjustment is instantaneous. Consequently, any disturbance to the system will immediately be absorbed and a new equilibrium will be established. It is as if all transactions occurred in a single market place, with a market organizer (auctioneer) determining prices on the basis of full market information and instantaneously announcing the going rates. The idea of price-related supply schedules and demand schedules applies to products and also to the factors of production (the rate of interest for capital and the wage rate for labour). On these assumptions, should there be an excess supply of goods, services, labour or capital, there would be a downward adjustment in the relevant price(s). This would clear the market. In addition, future supplies would be less abundant, so that an equilibrium between supply and demand would be maintained. Conversely, a deficient supply will stimulate higher prices, lower consumption and an increase in the supply offered, leading to a new equilibrium position. Embedded in this structure of thought is Say's Law. Colloquially, this states that 'supply creates its own demand'. A more formal definition of the law is given by Gilpin (1973, p. 189):

> every increase in the supply of goods is [matched by] a corresponding increase in the demand for goods so long as producers have directed their production in accordance with each other's wants.

The proviso is obviously of crucial importance. If it is not met, unwanted goods might accumulate and the circular flow of supply-demand-supply would grind to a halt. Potentially of equal seriousness is the possibility of delay between the receipt of income and its expenditure on goods and services. Such delay could vitiate an alternative version of Say's Law, which holds that the expenditure incurred in production releases into the economy purchasing power which, in aggregate, is sufficient to ensure that all of the initial output is consumed.

On the assumption that markets for the factors of production, and for goods and services, respond instantaneously to market signals, and that income is spent immediately, Say's Law must hold. By the same assumptions, though, it must be the case that an increase in demand will create its own supply. In both cases, the limiting situation would be given by the condition of full employment, when all the labour resources are utilised to the full.

Writing in 1965, Stigler (p. 256) had this to say about the idea of perfect competition, which provided the foundation for neo-classical economic doctrine:

> The concept of perfect competition received its complete formulation in Frank Knight's *Risk, Uncertainty and Profit* (1921). It was the meticulous discussion in this work that did most to drive home to economists generally the austere nature of the rigorously defined concept and so prepared the way for the widespread reaction against it in the 1930s.

This reaction took two directions of relevance to us, one working within the neo-classical tradition and the other turning that tradition on its head – imperfect and monopolistic competition on the one hand, and Keynes on the other. But before we turn to these schools of thought, a word must be said about the nature of the conclusions regarding government policy which flowed from the austere concept of perfect competition.

The strongest conclusion is the proposition that market mechanisms provide the most efficient allocation of resources and that any intervention which 'distorts' the system away from the market optimality is undesirable. If this conclusion is accepted, then governments ought not to intervene at all, except in so far that taxation and expenditure are necessary in order to provide collective goods and services, such as defence against external enemies and the maintenance of internal public order. Alternatively, if it can be shown that markets operate less than perfectly, then the role of government is to intervene to remove the imperfections – to limit the emergence of monopolies and cartels, to ensure the easy flow of information, to control corruption, etc. A government would be in a position analogous to that of an umpire, charged to ensure that clearly defined rules are administered strictly but in an even-handed manner. Regional problems are given no prominence in this scheme of things. Chronic disequilibrium, manifest in persistently high levels of unemployment, should not occur. If it does, then the interpretation emphasizes the failure of capital and/or labour to adjust adequately, so that a 'regional' problem is perceived to be no more than the spatial manifestation of an adjustment failure by the factors of production, which leads to an examination of measures to facilitate the requisite adjustment.

## Monopolistic competition

The year 1933 was remarkable for the publication of four major books, written by scholars from four nations. The works of Chamberlin and Robinson mounted a strong attack on the neo-classical doctrines pertaining to an imagined world of perfect competition. Both explicitly drew attention to the spatial implications of monopolistic and imperfect competition, and to the fact that the costs of overcoming space provide an important reason why no real economy can function as postulated by the perfect competition model. This point was underscored by Ohlin's study of international and interregional trade, in which the existence of transfer costs must be accepted as a central fact, modifying the conclusions derived by assuming zero costs of movement. Together, these three studies marked a major departure in economic thinking, opening the door to a concern with spatial organization. Coincidentally, Christaller also published in 1933 his highly original study of central places, based on the principle of spatial monopolies nesting in a hierarchy.

In the spatial domain, two important publications had appeared just four years earlier. Weber's 1909 study of industrial location, originally published in German, appeared in English in 1929. Focusing on transport costs, Weber worked within a strong nineteenth-century tradition among German scholars; the translation of his study brought that tradition squarely into focus. Finally, also in 1929, the *Economic Journal* published Hotelling's famous paper concerned with the price and/or location strategies of firms in a situation of duopoly.

Although 'monopolistic' and 'imperfect' competition should probably be regarded as distinct from each other and certainly the books by Chamberlin and Robinson are very different – nevertheless there are important common elements. Both mount a strong critique of neo-classical economic doctrine. Robinson approached this in part from the debate which was then ongoing about the significance of economies of scale, the existence of which undermines assumptions in the model of perfect competition, and in part from the 1930s experience of worldwide slump. In the preface to the second edition, she summarizes the Pigovian version of Marshallian doctrine, and then continues:

Here we were, in 1930, in a deep slump, and this is what we were being asked to believe ... The first point in Pigou's scheme was patently absurd. Under perfect competition, any plant that was working at all must be working up to capacity ... Imperfect competition came in to explain the fact, in the world around us, that more or less all plants were working part time.

(Robinson, 1969 edition, p. vi)

Instead of treating perfect competition as the norm, with monopoly as a special case, Robinson concludes that monopoly is the normal situation. In contrast, Chamberlin's critique, arising from work started in 1924 and completed in 1927, focused on the more general problem of how to integrate perfect competition with monopoly, duopoly and oligopoly and simultaneously to reflect some elements of the reality of economic systems. His central position is that all situations embody some degree of competition, if only the competition between substitutes, and also some elements of monopoly, such as that conferred by product differentiation and/or geographical location.

For both authors, the central problem was to determine value, price and output in a world other than that of perfect competition. The subtitle to Chamberlin's book, 'A re-orientation of the theory of value', conveys this message. In such an undertaking, one has to consider the conditions which will permit equilibrium to be achieved. Thus, although monopolistic competition theory (and imperfect competition) severely criticizes neo-classical thought, it is itself a corpus of ideas which is firmly *within* the tradition it criticizes. That is to say that both Chamberlin and Robinson were thinking in micro-economic terms, in a search for equilibrium, but with concern for a theory which would bear some resemblance to the real world. This frame of reference, taking technology as given and neglecting macro-economic issues, is essentially static in character.

Under monopolistic competition, a firm faces a situation which differs radically from that postulated in perfect competition theory. For our purpose, two differences are particularly noteworthy. If, initially, we assume that conditions of perfect competition exist, there will be an homogeneous product, for which demand can be represented by AD in Figure 3.1. If the supply available from the very large number of producers is AS, the total output will be OQ and the ruling market price will be OP. Because each producer contributes only a very small (negligible) proportion of the output, he can have no influence on the ruling market price. Consequently, the horizontal line PX represents the demand curve which faces each producer: whatever he can produce he can sell at that price. In the perfect competition model, therefore, the demand curve facing each producer is conceived to be horizontal.

Were there to be a single producer, he would be faced with the same aggregate demand curve (AD), sloping downward to the right as shown in Figure 3.1. However, the monopolist operates in a market which is fundamentally different from that of the firm in a perfectly competitive world. A monopolist can contemplate deliberately curtailing output in order to raise the market price, if he judges that this will maximize his profits. Conventional

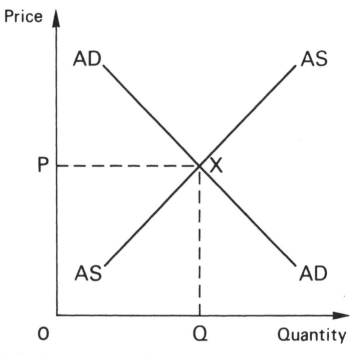

**Figure 3.1** Supply-demand relationships

formal analysis indicates that, to achieve maximum profits, a monopolist will indeed restrict output. Should there be two firms (duopoly) or several (oligopoly), the aggregate demand curve may still be taken to represent the demand curve facing each firm. But as the number of firms increases, so will the demand curve facing each increasingly approximate to the horizontal. The general point to be made is that, under conditions of monopolistic competition, the demand curve can be assumed to slope to the right in some degree, rather than being horizontal.

Very few products, if any, are truly homogeneous. Natural products, such as coal and timber, possess widely varying characteristics which render certain sources suitable for particular purposes. Manufactured products are every bit as diverse, and may be deliberately fashioned to serve particular functional needs (compare a family saloon car and a Landrover) or to cater for personal whim and fancy, as with fashion clothing. Manufacturers are constantly striving to develop and market products for which there is no exact competition, and thereby to create an element of monopoly. Most markets can be characterized as competition between products which are partial substitutes for each other. In this way, most firms have, or seek to create, an element of monopoly by differentiating their product from that of their rivals. Such differentiation may be supported by patents and trademarks. Alternatively, differentiation may rest largely on consumer preferences conditioned by advertisements.

Market segmentation is a basic characteristic of monopolistic competition.

Such segmentation may occur for either or both of the following reasons: product differentiation, as discussed above; and spatial monopolies, i.e., the existence of discrete market areas. In both cases, the demand curve facing the seller falls to the right. In both cases, therefore, the price to the consumer will have some influence on the quantity purchased, depending upon the shape of the demand curve and the price elasticity of demand. In the spatial case, it is commonly assumed that the customer bears the transport costs of getting his purchases home. The 'delivered' price will therefore vary spatially around the point of supply, with the implication that consumers move *up* their individual demand curves, purchasing less and less, at increasing distances (Figure 3.2). Where there are two suppliers, located at $0_1$ and $0_2$, it is usual to assume that customers will patronize the outlet which gives them the lowest delivered price (in Figure 3.2, X indicates the point of indifference). On this assumption, discrete market areas can be generated.

The perfect competition model makes some strong assumptions regarding economies of scale for individual firms. If unit costs of production are constant, then every firm competing in a market for an homogeneous product could expand its output to supply the full quantity being demanded; a single supplier would suffice. Conversely, an infinitely small output would also be viable, giving an infinite number of firms. To cope with the problems posed by these possibilities, it became habitual to specify that each firm is confronted by a U-shaped cost curve, and that the point of inflection is the only point at which a firm could earn the minimum necessary profit. Any departure from that point would result in lower profits being earned, with the result that the firm would go out of business. These assumptions are unrealistic. In the monpolistic competition model, scale economies become a more important consideration. First,

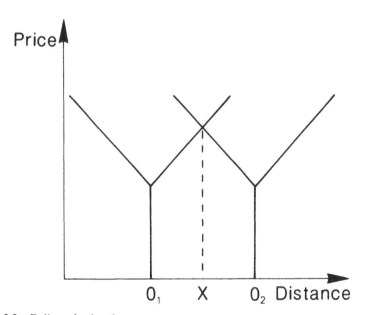

**Figure 3.2** Delivered prices in space

the shape of the cost curve varies from product to product (firm to firm), and may be a powerful reason why the number of firms making a particular commodity is limited to one or a small number. Second, the combination of scale economies with the options firms have for price and/or location strategies opens up some important further questions concerning their behaviour. With discrete spatial markets, the size of the market area will affect total sales, total costs of transfer and also the unit costs of manufacture.

The existence of monopolistic competition implies that there will be an element of uncertainty in the economic system. In taking their decisions, entrepreneurs must consider the probable responses of their competitors – responses which may be locational, may involve product differentiation or may relate only to price. If, and only if, firms are able to make accurate predictions of their rivals' responses will it be possible to conceive of a stable equilibrium for firms in competition.

Sufficient has been said to show just how radical was the critique of neo-classical economics mounted by Robinson and, especially, Chamberlin. It will also be appreciated that to formulate monopolistic competition theory in a logically consistent way, to establish the conditions necessary for equilibrium, involves complex analysis. One of the leading scholars to essay this task, and to do so with the spatial dimension explicitly in mind, has been Greenhut (1956, 1963 and 1970; Greenhut and Ohta, 1975; Greenhut *et al.*, 1987).

The existence of monopolistic competition in space invites the question: Can rules be specified which will yield a stable pattern of locations under conditions of duopoly and oligopoly? This is the question which Hotelling asked in 1929 and is a question on which there has been considerable debate; this debate has been thoroughly examined by Greenhut (1970). Furthermore, can an equilibrium for firms be generalized to the whole space economy? In other words, was Christaller's exploration of central places something which could be rigorously and fruitfully examined in the light of monopolistic competition theory? This intriguing question was tackled by Lösch (1944, English translation 1954) and Isard (1956). The appearance in 1954 of an English translation of Lösch's major study, following as it did, the appearance of Dickinson's 1947 study of cities and regions, Hoover's 1948 examination of the location of economic activities and Zipf's 1949 treatise on human behaviour viewed in terms of effort minimization, created a receptive audience for Isard's 1956 study.

To view the space-economy in general terms is to think of the interdependence of locations and activities, the link between location and price and the role of transport costs. Isard formulated the last issue as a substitution problem – the actual or potential substitution of transport for other inputs, and vice versa. To handle the multiplicity of linkages, one must conceive the space-economy as structured by a set (a large set!) of simultaneous equations. The central question then becomes the following: Can one specify the most efficient configuration of the space-economy, and will that be stable? In other words, can a spatial equilibrium be visualized? Lösch answered that question in the affirmative. Writing of the equilibrium of locations, he remarked:

> This is determined by two fundamental tendencies: the tendency as seen from the standpoint of the individual firm and hitherto alone considered, to the maximization of advantages; and, as seen from the standpoint of the economy as a whole, the tendency to maximization of the number of independent economic units. The latter is affected by competition from without, the former by industrial struggle within. The individual chooses his location in such a way as to achieve the highest profit as a producer, or the cheapest market as a consumer. But in so doing, as though it were a trick of the idea, he makes possible the existence of more competitors. They crowd into the market and reduce his living space until his advantage disappears. There is constant struggle between two forces; what is gained by the one is taken back again by the other.
>
> The point where these forces balance determines location. This equilibrium, born of the interdependence of locations, can be understood only through a system of general equations of location. As soon as the conditions expressed by these equations have been fulfilled the struggle for space dies down, and when the equations are solved the locations themselves are determined.
>
> We now present the general conditions of equilibrium that are valid for independent producers and consumers, for agriculture as well as for industry, and develop the *pertinent* equations briefly for the latter.
>
> (Lösch, 1954, p. 94)

In order to write such a set of equations, one has to be able to specify the behaviour rules for firms and for consumers, i.e., it must be possible to cope with uncertainty. Furthermore, one has to consider how to resolve the potential (or actual) conflict between the outcome which is optimal for the firms and that which is optimal from the viewpoint of consumers. Even if these two conditions can be met, the data requirements are immense if real-world situtations are to be analysed.

Christaller's basic insight (see Figure 3.3) stood up remarkably well to the subsequent scrutiny. Both Lösch and Isard substantially modified his ideas for the hierarchy of settlements but both had to remain content with a very general and schematic representation of the ideal space economy which they envisaged (e.g., Isard, 1956, Figure 52). The achievement was to have systematized the conditions required for an efficient ordering of the economic landscape, based on the principle of minimizing total movement costs, subject to contraints, such as the distribution of the population engaged in primary production. These ideas have been influential in the planning of individual settlements and in some cases whole settlement systems; a notable example is the planned pattern of towns and villages in the polders which the Dutch have reclaimed from the sea. The general equilibrium approach to the space economy allows one to examine the efficiency of spatial structures, given the total activity to be distributed in

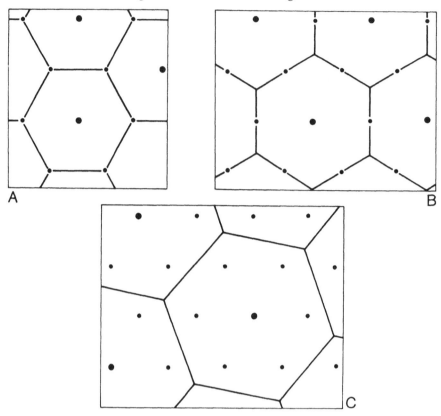

**Figure 3.3**  Central place systems after Christaller
A κ = 3      B κ = 4      C κ = 7
*Source*: Haggett, 1965, p. 119

space and such constraints as may be specified. Once the bricks and mortar which give spatial structure its expression are in place, they will materially influence the future allocation of increments (or decrements) to the space economy. But, as an analytical frame of analysis, the focus is on the efficient *allocation* of resources spatially, not on the reasons why resources grow and decline.

In his study of industrial location, Weber (1929) took transport cost minimization as the overriding regional organizing principle. He also very explicitly regarded labour costs as a major factor and, at the subregional scale, elaborated the concept of scale economies, both internal and external to the firm. The latter formed part of his 'agglomeration' effect, which might cause firms to diverge from their transport optimum location. These ideas were strongly reinforced by the appearance in 1931 of Austin Robinson's study of the structure of firms in competition, in which he discusses the sources of scale economies and the concept of optimum firm size. In addition, Young (1928) published his influential study of the sources of national economic growth, in which he argued that increasing returns are obtained from the realization of scale economies. In the context of Joan Robinson's ideas and those of Chamberlin, the scene was set to accord economies of scale an important place in the economic system, instead of the minor (dare one say marginal?) place accorded thereto in neo-classical thought.

It is but a short step to allow the following possibility. If some change occurs in the space economy, from whatever cause, opportunities may be created for existing firms to realize further economies of scale, or for a new firm to be formed to carry out tasks formerly undertaken within existing businesses. The realization of these scale economies may further change the interrelationship of firms, to allow yet more economies to be realized. Such a process will occur if the localization of one firm implies the creation of infrastructure facilities which have surplus capacity and thereby create opportunities for other firms. In addition, the range over which scale economies operate may differ from one activity to another. To obtain the full benefits of scale economies in the provision of services such as electricity supply or transport provision will require a city much larger than is needed for the optimal working of a barber's shop.

Isard, following Weber and Hoover in particular, devoted a considerable

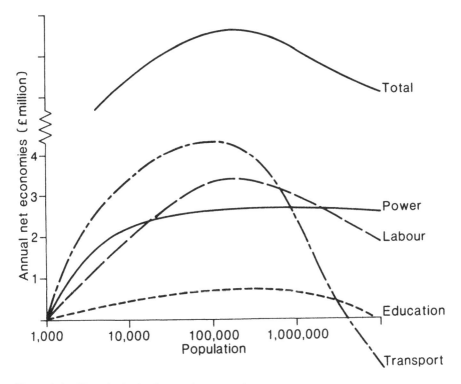

**Figure 3.4** Hypothetical urban scale economies
*Source*: Isard, 1956, p. 187

amount of space to the implications of scale economies for the spatial organization of activities. Figure 3.4 reproduces his illustrative diagram for urban scale economies. At this juncture, there are three points to make regarding the significance of scale economies for the geography of economic activities.

(1)   Internal economies of scale tend to reduce the number of firms needed to supply a given total demand, thereby reducing the number in competition. External scale economies will cause firms to agglomerate to an extent much greater than would otherwise to the case.

(2)   To the extent that there are unrealized economies of scale, and especially of external economies, there is in the space economy a source of dynamic change which will involve spatial adjustments. These will appear locally as growth (or decline), and the realization of these economies will allow aggregate output in the whole system to rise, at a given level of resource use. An adjustment of this kind may occur quickly, or, if there are substantial imperfections in the system, may continue over an extended period.

(3)   Economies of scale will again come to our attention in the context of regional multiplier analysis.

In his analysis of monopolistic competition, Chamberlin refers to the problems posed if firms are uncertain about the future, and especially concerning the responses of other firms to any change they may make in the price for their product or the quantity marketed. This particular problem had been carefully explored by Knight (1921), who drew the distinction between 'risk' and 'uncertainty'. The former term describes situations in which the probabilities of outcomes can be accurately measured, such that one may speak of 'objective probability'. Uncertainty, on the other hand, denotes a situation in which the probabilities cannot be specified precisely; a situation of 'subjective probability'. Although Knight recognized that if there is uncertainty an economic system cannot attain perfect equilibrium, he was looking for the ways in which the degree of uncertainty could be minimized. In particular, he examined the strategies for converting uncertainty into risk, or something approximating risk, whereby the system could again be determinate. Chamberlin seems to have accepted this line of thought. Consequently, his treatment of uncertainty is fairly brief, and in some senses tangential to his main concern.

## The Keynesian revolution

According to the accepted wisdom of the neo-classical school of thought, any reduction in national output and the associated increase in unemployment would be a temporary phenomenon, because the self-correcting mechanisms of lower wages, prices and interest rates would ensure the restoration of the economy to its (labour) supply-determined level of full employment. This view of events appeared to fit the cyclical behaviour of economies during the nineteenth century and the first quarter of the twentieth. However, the recession of the early 1930s was on a scale and of a duration that far exceeded previous experience. Despite the fact that wages fell, that the prices of goods declined and that interest rates dropped, the self-correcting mechanisms manifestly were inadequate to return economies to a high level of activity within a reasonable period of time. The scale of the unemployment problem, the duration of the enforced idleness, and the hardship associated therewith, were unacceptable. In other words, a catastrophic event occurred which, according to the then orthodox opinion, was 'impossible'.

In writings that culminated with his 1936 *General Theory*, Keynes offered a radically different insight into the functioning of modern economies. As Deane observed:

> The really shattering conclusion to emerge from the *General Theory* in the context of the 1930s was that a cut in wages, far from relieving unemployment as the classical theory implied, could actually increase it by reducing the level of effective demand ... [Keynes] demonstrated in a way that had never been so dramatically and clearly brought out before, the dependence of aggregate expenditure on itself in its income-generating capacity.
>
> (Deane, 1978, p. 183)

Thereby, Keynes showed that there is no necessity for an economy to move to an equilibrium position in which labour resources would be fully used. Whether or not an equilibrium would be achieved, high levels of unemployment could be a permanent or quasi-permanent condition. Equally important, though, Keynes provided an analytical structure which showed that government intervention to stimulate aggregate demand could be expected to maintain an economy in a state of full employment and, by extension, return it to that state from a condition of recession.

Keynes was a prolific writer, and the literature which has sought to interpret or to attack his ideas is positively vast. If one may use Dow's (1985) term, there is a large amount of 'product differentiation', as scholars seek to establish their own stall in the market place of ideas. We will follow Coddington (1983) and take a relaxed view of the terminological jungle. We will use the term 'Keynesian' to denote the analytical mode of thought which has a clear ancestry in the work of Keynes, which became the conventional wisdom of the 1950s and 1960s, but without wishing to imply that Keynes himself either held to particular points or would subscribe to them were he alive today. In other words, we will present a highly condensed and conflated version of Keynesian thinking that can be contrasted with neo-classical thought, relevant for the main theme of the present book.

However, before we embark on this account, we should note the existence in Keynes' own thinking of two quite distinct strands:

> after twenty years' meditation on Keynes and his system, the writer has been forced to conclude that most of the confusions connected with the system spring from the fact that Keynes was fundamentally torn, though perhaps only semiconsciously so, between two quite opposing basic ideas

of the nature of the economic world. On the one hand he had a very pedestrian, mechanical view of economics, based on the assumed existence of a simple, direct, cause and effect mechanism. On the other side he had a highly subjective, futuristic approach in which the cause and effect relations are far less definite and inevitable.

(Wright, 1962, pp. 48–9)

It was the 'mechanical' view of the economy which, codified by Samuelson, forms the basis of what we know as Keynesian economics, even though this almost certainly misrepresents his ideas on many matters (Meltzer, 1989).

Why it should have been this part of Keynes' work which attracted the most attention is a matter for historians. In the British context, wartime experience suggested to some, including men as influential as Beveridge, that Keynesian policies did work and could provide the basis for managing the peacetime economy in a way that would make it possible to build a New Jerusalem immediately (Pollard, 1982). Then, during the 1950s and 1960s, the ideas of Hempel and Popper came to have important influence among social scientists, and the Keynesian economics then in vogue appeared to provide the basis for a true economic science. Yet, as early as 1954, Schumpeter commented of Keynes and his *General Theory*:

his work is a striking example of what we have called above the Ricardian Vice, namely, the habit of piling a heavy load of practical conclusions upon a tenuous groundwork, which was unequal to it yet seemed in its simplicity not only attractive but also convincing.

(Schumpeter, 1954, p. 1171)

In contrast to the micro-economic focus of neo-classical thinking, Keynes directed attention to the performance of the economy in aggregate terms. Much of the relevant groundwork had been laid in the 1920s and 1930s, on national income accounts at both the theoretical and the empirical levels. On the assumption that in the short run the size of the workforce is constant, total employment will be some function of total output, and, on a first approximation, unemployment will be the difference between the total number of people seeking work and the number of jobs available. Therefore, if the number unemployed is to be regulated, the number in work must be controlled, and this in turn requires that government be able purposively to affect the level of output, or the Gross National Product (GNP). In pursuit of this aim, Keynes visualized the circular flow of funds within the economy in terms that may be represented as follows:

$$Y = C + I + G + E = C + S + T + M$$

| where | | | | |
|---|---|---|---|---|
| | Y | = Gross National Product | C | = Consumption |
| | I | = Investment | S | = Savings |
| | G | = Government expenditure | T | = Taxes |
| | E | = Exports | M | = Imports |

The above identity must, by definition, be true for past levels of Y. However, for future states of the economy a crucial distinction should be made between expected behaviour (e.g., with respect to investment decisions) and the actual outcome, such that a stable level (or rate of change) of Y depends on realized outcomes being equal to expected outcomes. For example, should actual consumption (C) or investment (I) be less than expected, deflationary pressures will appear in the economy, leading to a lower level of GNP than had been expected.

The central problem on which Keynesian thinking focuses is the following. Given that in the short run the total population is constant, the level of unemployment will be determined by the number of people in work. The number employed will be a function of Y, or total output. If it can be assumed that there is a one-to-one relationship between output and employment, the problem becomes one of managing the economy to obtain a level of Y which will give the desired level of employment and hence of unemployment (recognizing that some unemployment is inevitable, if only the frictional unemployment of those changing jobs). Any unemployment which exceeds this basic minimum must, in this scheme of things, be due to a deficiency of demand. Therefore, the question reduces to identifying the ways in which government may manipulate aggregate demand to achieve the desired value for Y. The key variables are government expenditure (G), the aggregate tax impost (T) and the effects that this has on consumption (C), and the role of interest rates as a regulator of investment (I). As a consequence, Keynes himself advocated the following tripartite package of policies to counteract recession:

(1)  Action to drive interest rates (I) down to stimulate investment.
(2)  Redistributive taxation (T), on the principle that people on lower incomes have a greater propensity to spend than those on higher incomes.
(3)  Large-scale public expenditure (G), to pump demand into the economy.

Thinking about the economy in the above terms involves the use of aggregate values and the assumption that these aggregates are themselves homogeneous. For example, it is assumed to be immaterial whether consumption (C) increases on account of higher expenditure on holidays, hospital care or humbugs; it is also unimportant whether total demand increases on account of higher consumption (C) or a larger volume of investment (I), since they will contribute to GNP (Y) in equal measure. These assumptions are obviously heroic. However, there are other assumptions which are equally important, of which the short period under consideration is perhaps the most notable. The Keynesian system depends on the constancy of relationships between the variables, and such constancy can only realistically be assumed for a comparatively short period. As Wright put the matter in commenting on a passage from the *General Theory*:

Putting this paragraph in simpler language it will be seen that it assumes that (1) there is no technical change or invention; (2) there is no change in taste; (3) there is no change in population or resources; and (4) there are no changes in the preferences of the population between work and goods on the one hand and leisure on the other. These assumptions in effect 'freeze' the system, and practically every dynamic element of capitalist civilization is removed.

> (Wright, 1962, p. 2)

In the Keynesian scheme of things, the economy is conceived to be demand driven, the role of government being to manipulate total demand to achieve the requisite level for the desired level of employment. This implies that output can and does respond promptly to changes in aggregate demand. Therefore, starting from a position at less than full employment, it is assumed that an increase in demand will call forth higher output without generating inflationary pressures. Put colloquially, demand will generate its own supply, which is to turn Say's Law on its head. Thus:

> The Keynesian dichotomy, in its basic, unqualified form, may be stated as the principle that output is determined by aggregate demand, and that prices are determined by costs.
>
> (Coddington, 1983, p. 11)

This view was modified with the publication of Phillips' (1958) study of the trade-off between inflation and unemployment. According to his analysis, at levels of unemployment below 2.5 per cent inflation is apt to be uncomfortably rapid. Unemployment at 2.5 per cent is about half the level which, during and immediately after the Second World War, had come to be regarded as acceptable. However, the validity of Phillips' analysis has been challenged, most notably in respect of how to incorporate workers' expectations into the wage bargaining process (Phelps, 1967) and the problem of how to specify the model (Johnes and Hyclak, 1989).

One of the crucial points on which Keynes departed from neo-classical thinking was the matter of wages – the price of labour – during a recession. According to neo-classical ideas, an excess of labour, as manifested by unemployment above seasonal and frictional norms, would drive down its price – wages – until the labour market was cleared and full employment restored. Keynesian economists take the view, in contrast, that contractual wage bargains ensure considerable downward stickiness of wages, so that in practice wage rates are slow to respond (see p. 124). In any case, because many other prices are also sticky, a downward movement of wages would likely lead to a reduction in total purchasing power and hence in aggregate demand, which would compound the problem that lower wages were supposed to solve. Therefore, a feature of central importance in Keynesian thinking is the assumption that there is considerable downward rigidity in real wages. To the extent that lower real wages are necessary, they should be achieved by the maintenance of nominal wages as employment picks up and there is upward pressure on prices. Thus, in conditions of excess supply of labour, Keynesian doctrine holds that the volume of employment 'is determined by the demand for it, and [is] impervious to the supply' (Coddington, 1983, p. 34).

It was also assumed that the prices of manufactured goods are generally sticky in a downward direction. Thus, an important element of Keynesian thinking is a denial of the central role of price changes for effecting adjustment, as postulated by neo-classical scholars. Consequently, Keynesian thinking emphasizes the role of quantitative changes – especially in output and employment – and plays down the significance of price adjustments (McCombie, 1988a).

Until the Great Depression and the appearance of *The General Theory* in 1936, it had been customary to recognize frictional and seasonal unemployment as 'acceptable' categories, and to regard any additional unemployment (of a cyclical nature in the generality of cases) as essentially voluntary in character. Keynes drove a coach and horses through the idea that large-scale unemployment could be voluntary. For him, massive numbers out of work implied involuntary idleness that was to be taken as a symptom of a deficiency of aggregate demand. Thus it is that the term 'demand deficient unemployment' came into common currency. But, as Coddington observes:

> If demand is deficient, it must be deficient in relation to something. That something is presumably the level of costs. So, with greater strictness, what we are talking about is 'a deficiency of demand in relation to costs', which can be thought of quite equivalently as an excess of costs in relation to demand.

> (Coddington, 1983, p. 41)

Lipsey (1968) pointed out that if there is a negatively sloping Phillips curve, then in principle there is no limit to the amount of unemployment that can be removed by expanding demand and accepting the consequential inflation. However, the marginal cost in terms of inflation will rise and may become unacceptable before unemployment has been completely eliminated. Therefore, 'demand deficient' unemployment is really only that part of total unemployment which can be removed more efficiently (more 'cheaply') by demand expansion than by other means, i.e., through the lowering of costs. This distinction may seem rather trivial. However, the term 'demand deficient' is highly loaded, and leads one to look for only one set of mechanisms on the demand side.

The last feature of the Keynesian system which needs to be considered is the role of multipliers, which are of central importance for the management of the economy; they also point the way to the analysis of regional economies. If a government is to manage the level of demand so that total output can be maintained at a specific level, it is necessary to know what ramifying effects there will be from a given initial change in demand. Suppose that expenditure initially increases by £1.0 million. That expenditure represents income for those who have sold goods and services, and they will have more money to spend. If they save one-tenth of their income – the propensity to save (m) equals 0.1 – the next round of expenditure will be only £0.9 million. The ripples of expenditure diminish with successive rounds. The overall multiplier effect is given by $1/m$, which in this hypothetical case is $1/0.1 = 10$. If sensible judgements are to be made concerning the magnitude of initial change in demand that is required in a given situation, it is essential to know what the relevant multiplier is. Otherwise, it is virtually certain that the effect of a change will be either to undershoot or overshoot the target. Consequently, much work has been devoted to the estimation of multipliers as part of the general programme of refining national income accounts and the art of econometric forecasting.

One of the attractions of the Keynesian vision was its treatment of demand as an endogenous and controllable element of a national economy. Although exports and imports are integral parts of the national income identity (p. 42), their significance in economic management was given relatively little prominence. Indeed, in order to manage demand effectively, it was assumed that the economy could be treated as if it were closed. Hence, it was supposed that foreign trade and the balance of payments were not problematic. In practice, they proved to be very awkward aspects of economic management. Thus, in an influential paper in 1971, Kaldor argued that governments should shift the focus of their demand management efforts away from domestic final consumption and toward exports, with exchange rate policy a key factor for the competitiveness of goods sent overseas. Kaldor was responding to the problems which had become manifest in managing an economy as if it were closed when in practice it is very open. His solution was to attempt to control aggregate demand through the volume of export business. Such an approach could only work if the major trading partners took no countervailing action to protect their own interests.

During the post-war period into the 1970s, Keynesian doctrines dominated economic thinking. Not surprisingly, major components of this tradition were incorporated into the work on regional economies that was in hand during this period. Although we will explore this theme in detail in Chapter 4, at this stage the following points deserve special mention. First, the Keynesian system is demand driven, with supply-side considerations given relatively little weight. Translated into regional terms, demand came to be conceived as exogenous to the region in question. Or, if extra demand were injected into a region through government intervention, this again was an action exogenous to the region. Finally, the impact of demand changes could be analysed via the role of multipliers, to estimate the overall change in employment and/or income.

## Radical theory

Neo-classical theory assumes that economic systems tend toward equilibrium and that this equilibrium is benign, giving harmonious relations between the workers and employers. Workers will be fully employed and earning wages related to their productivity, while employers reap 'normal' profits. The Keynesian system, on the one hand, and monopolistic competition theory on the other, are denials of this comfortable view of the world. Keynes showed that equilibrium could be far from benign, with chronic unemployment in the absence of government intervention, but took the optimistic view that government could intervene to achieve a situation satisfactory for employers and for workers. Monopolistic competition theory tends to the proposition that a firm which enjoys a monopoly can maximize its profits by manipulating price and

setting output below the maximum which is possible. The interests of consumers and producers diverge, and the greater the element of monopoly, the
greater the ability of firms to gain advantage. Explicit regulation may be
necessary to ensure that the interests of consumers are protected.

The neo-Marxist critique goes a good deal further. It is asserted that the
technical relations of production and consumption can only be understood by
reference to the social relations, the two sets of relationship constituting a mode
of production (Cole, Cameron and Edwards, 1983). A binary distinction is
made between the owners of the means of production (capitalists) and those
who only have their labour to sell (workers). For the productive system to
remain in being from one period to the next, sufficient goods and services must
be produced so that everyone may live and reproduce. The first purpose of
economic activity, therefore, is to maintain the mode of production. Over and
above that necessary output, there will be a surplus, and it is the way in which
that surplus is distributed between the capitalists and the workers which
provides a major focus of concern.

The fundamental proposition seems to be that division of the surplus is a
zero-sum game and therefore a matter of conflict between the classes. As
Sheppard and Barnes (1986) put it, profits are inversely related to wages.
Furthermore, it is generally assumed that the capitalist class has the upper hand
and is therefore able to expropriate most, if not all, of the surplus, whereby the
working class is exploited. On this analysis, workers must organize themselves
so as to wrest a larger share of the surplus than they would otherwise obtain.

Class conflict over the division of the surplus is therefore seen as a central
feature of capitalist production and the driving force for change. In this
representation, the capitalist class is generally seen to be well organized and
able to act in the manner of a monopolist, such that the terms 'monopoly
capital' and 'state monopoly capitalism' are quite widely used (Damette, 1980;
Harvey, 1982; Semmler, 1984). In practice, the owners of capital do not
necessarily have congruent interests, a fact which may either generate conflict
within the capitalist class, and hence within capitalism, or at least lead to
divergent paths of development (Massey, 1984, p. 27). However, it is often
assumed that the capitalist class is homogeneous and well organized, whereas
the labouring class is regarded as fragmented, almost atomistic. This view is
illustrated by Harvey (1982, p. 384) in his discussion of the spatial implications
of Marxist analysis: 'Capital in general relies upon this perpetual search by
workers for a better life – defined in material and money terms – as means to
orchestrate labour mobility to its requirements and to discipline individual
capitalists to class requirements.' Note the personification of 'capital' acting as
the 'hidden hand' of classical and neo-classical economic thought.

Deeply embedded in the Marxist tradition is the belief that market prices
provide an insubstantial basis on which to build an economic theory. Marx
himself, and his followers, turn to the labour theory of value as the foundation
for analysis:

(i) to explain the equilibrium prices (or the exchange-values) of commodities, around which actual prices fluctuate over time, and (ii) to provide

aggregators, or weights of aggregation, in terms of which a large number of industries (or primitive sectors) are aggregated.

(Morishima, 1973, p. 10)

The belief is that a labour theory of value will yield an 'objective' or 'real' basis for economic accounting, which is devoid of the subjectivity inherent in market prices. Now it is immediately clear that value cannot be directly proportional to the amount of labour expended, since this would lead to the perverse proposition that an article which took two hours to make has twice the value of an identical article which a more dexterous worker can fashion in one hour with the same tools. To overcome this difficulty, a sophisticated system of relationships has been evolved which requires distinctions between 'use value', 'exchange value' and just 'value'. Labour itself must be reduced to the concept of 'abstract labour', and some notion of a living wage must be incorporated. As a result, even the most elementary exposition of Marxist economics very quickly becomes highly complex, the more so given that: 'the concept of class is embedded in the conception of value itself' (Harvey, 1982, p. 33).

One of the main conclusions to which Morishima (1973) came in the light of his exhaustive and sympathetic examination of Marx's economics is that the labour theory of value is in fact not satisfactory:

We conclude by suggesting to Marxian economists that they ought radically to change their attitude towards the labour theory of value. If it has to determine the amounts of labour which the techniques of production actually adopted in a capitalist economy require, directly or indirectly, in order to produce commodities, it is not a satisfactory theory at all. As has been shown above, the value system may be determined to be negative, indefinite or even contradictory to the postulate of the uniform rate of exploitation. These findings urge us to abandon the theory.

For a thorough-going Marxist it would be impossible to conceive of Marxian economics without the labour theory of value. Since it provides the workers with an inspiring ideological rationale for their struggles against bourgeois regimes, Marxists would be greatly depressed by losing its authority. In addition to this emotional damage, the foundations of Marxian economics, as a two-department macro-dynamic theory, would be seriously shaken.

(Morishima, 1973, pp. 193–4)

One of the central purposes of Marxian economics is to analyse the development path for capitalist economics. Much reliance is placed, in this context, upon the proposition that profits must inevitably fall, as the accumulation of capital outruns the possibilities for its profitable employment. Indeed, this postulated tendency is elevated to the role of axiom. Notwithstanding that the owners of capital seek to prevent this happening – by investing in higher productivity or switching the location of their investment to areas with lower operating costs (usually conceived as lower wage costs) – it is regarded as inevitable that there will be periodic crises of 'overaccumulation'. Such crises are otherwise known as severe recessions, when productive capacity and stocks in hand exceed current demand. Adjustment to this situation will entail the closure of productive capacity and an increase in unemployment. These recurrent crises are regarded as inevitable. In this sense, Marxian analysis contests the conclusion reached by Keynes, that judicious intervention by government can keep an economy operating at the full use of resources. Given the apparent success of Keynesian demand management policies during the 1950s and 1960s, the Marxian view of crisis as a recurring condition attracted relatively little attention. However, with the ending of the Keynesian consensus in face of 'stagflation' (p. 53) and the experience of severe recession in the 1970s, Marxist analysis has had something of a vogue.

Marx focused upon the struggle within a nation between labour and capital, to appropriate the surplus product. One may visualize a parallel struggle between nations, in which the richer nations grow richer at the expense of the poorer. The clearest statement of this viewpoint has been made by the underdevelopment school of economists, especially in the context of the idea of 'unequal exchange'. This corpus of ideas has roots in the work of Prebisch with the Economic Commission for Latin America; their influential *Economic Survey of Latin America, 1949* was published in 1951 (Lipietz, 1987, p. 64). The underdevelopment school of thought amounts to a denial and rejection of the optimistic view of economic growth advanced by Rostow in 1960, in his *The Stages of Economic Growth.*

In attempting to describe this school of thinking, there are some major problems arising from the admixture of traditions. On the one hand, Marx himself took the view that capitalist development was a good thing, in that he believed it would create the procondition for socialist revolution (Corbridge, 1986). On the other hand, Frank (1967) argues that the penetration of world capitalism into the countries we now call underdeveloped has been an unmitigated disaster, in that the present state of underdevelopment has been created by that very penetration. In arriving at that conclusion, Frank very clearly relies on certain ideas which are unmistakably part of the Marxist tradition.

The underdevelopment school of thought holds that the undoubted increase in the prosperity of the richer countries has been achieved at the expense of the poorer, which, through a process of immiseration, have become underdeveloped. This implicit zero-sum view of global development has been clearly articulated by Frank. Writing specifically of Chile, he says:

This essay contends that underdevelopment in Chile is the necessary product of four centuries of capitalist development and of the internal contradictions of capitalism itself. These contradictions are the expropriation of economic surplus from the many and its appropriation by the few, the polarization of the capitalist system into metropolitan center and peripheral satellites, and the continuity of the fundamental structure of the capitalist system throughout the history of its expansion and transformation, due to the persistence or re-creation of these contradictions everywhere and at all times. My thesis is that these capitalist contradictions and the historical development of the capitalist system have generated

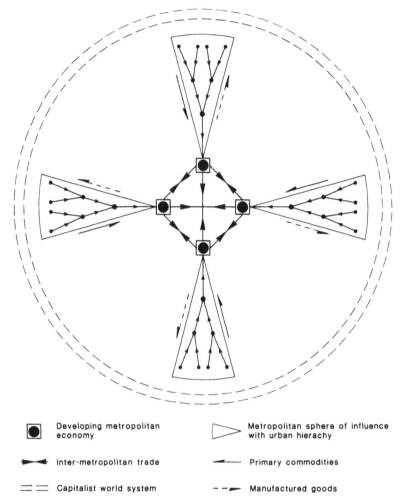

| | Developing metropolitan economy | | Metropolitan sphere of influence with urban hierachy |
| --- | --- | --- | --- |
| | Inter-metropolitan trade | | Primary commodities |
| | Capitalist world system | | Manufactured goods |

**Figure 3.5**   Resource transfers through the urban system, underdevelopment school of thought
*Source*: Corbridge, 1986, p. 31

underdevelopment in the peripheral satellites whose economic surplus was expropriated, while generating economic development in the metropolitan centers which appropriate that surplus – and, further, that this process still continues.

(Frank, 1967, p. 3)

Elsewhere, it is abundantly clear that he views all underdeveloped countries as experiencing the same outward transfer of 'economic surplus' through the urban hierarchy, in the manner indicated in Figure 3.5.

According to Edwards (1985), it was Prebisch who suggested that shifts in the terms of trade provided a mechanism for the systematic transfer of resources from poorer to richer countries. Over the period from the 1870s to 1939, the barter terms of trade for the United Kingdom moved in favour of manufactures and against primary goods and it appeared that this shift might be a general feature of the world economy. Although the evidence is in fact conflicting (see, for example, Chisholm, 1982, and Edwards, 1985), Prebisch paved the way for a more general concept, known as 'unequal exchange', which was propounded by Amin (1974) and Emmanuel (1972) in particular.

Traditional theorizing about international trade, from Ricardo through the Hekscher-Ohlin factor proportions theorem, holds that, when two or more countries engage in trade, the terms on which that trade occurs will be located within certain limits. Any ratio of imports to exports which is within those limits will leave both (all) trade partners better off than would be the case were no trade to occur, although the gains may be unequally distributed. If the trade ratios are at one extreme or another, then one (or more) countries will gain the entire advantage and the other (others) will gain nothing. On the assumptions normally made in trade theory, no trade will occur outside the limiting ratios.

Unequal exchange is the most general mechanism advanced to account for the systematic transfer of resources from poorer to richer countries, as portrayed in Figure 3.5. In essence, it is proposed that trade occurs on terms which lie outwith the limiting ratios of conventional theory. One of the more forceful protagonists of the unequal exchange thesis is Amin, who begins his discussion with severe criticism of neo-classical trade theory, which he characterizes as allowing:

> itself to make whatever assumptions it likes (assumptions that conflict with the facts) and thus to become a mere *jeu d'esprit* that refuses to take account of the facts; and this degeneration, due to its function as an apologetic ideology of universal harmonies, is closely linked with the subjective theory of value.
>
> (Amin, 1974, p. 43)

A fundamental objection to the mainstream theory is that: 'every change in the movement of commodities alters the comparative advantages because it affects the relative prices of the factors' (Amin, 1974, p. 51). If this proposition is accepted, there is no permanence to the patterns of comparative advantage and the analyst is caught up with the problems of subjectivity embodied in any market price.

The formal reasoning which underpins the unequal exchange thesis is complex; the reader will find the arguments summarized, and criticized, in the Appendix on p. 183. The labour theory of value provides the basis, in an attempt to explore the 'real' situation which lies behind appearances. It is believed that thereby the cancer of subjectivity, implicit in the use of market prices, can be avoided. However, as the Appendix shows, the doctrine of unequal exchange itself involves subjective judgements and heroic assumptions which are every bit as serious as those of which Amin complained.

Empirical evidence to substantiate the long-term net transfer of resources

from poorer to richer countries is scarce. Amin himself cites but one example.
In 1966, the underdeveloped countries exported a total of $35 billion, of which
$26 billion originated from the modern sector (minerals, oil, etc.) and $9 billion
from the traditional sector. Amin argues that if these commodities had all been
produced in the advanced nations at the prevailing level of productivity and
reward for labour, they would have had a collective value of $57 billion. On this
basis:

> if exports from the periphery amount to about $35 billion, their value, if
> the rewards of labor were equivalent to what they are at the center, *with
> equal productivity*, would be about $57 billion. The hidden transfers of
> value from the periphery to the center, due to the mechanisms of unequal
> exchange, are of the order of $22 billion.
>
> (Amin, 1974, pp. 58–9)

This conclusion is followed by this additional statement:

> exports from the underdeveloped countries represent 20 per cent of *their*
> product – about $150 billion. The hidden transfer of value due to unequal
> exchange is thus around 15 per cent of this product; this is far from
> negligible in relative terms, and is alone sufficient to account for the
> blocking of the growth of the periphery and the increasing gap between it
> and the center.
>
> (Amin, 1974, p. 59)

Two problems arise with this particular presentation of the unequal exchange
thesis. The first is that an actual market price of $35 billion is compared with an
imputed value of $57 billion. Such a comparison seems to suffer from all of the
defects Amin attributes to neo-classical 'pseudo-theory'. In the second place,
were all the production exported from the underdeveloped countries to have
been produced in the developed nations, what would the impact have been on
relative prices and quantities? To treat the 'hidden' transfer as if it were a 'real'
one does not seem to advance the argument at all.

Sheppard and Barnes (1990) review both the theory of unequal exchange and
its empirical applications. They identify only four empirical studies, not including Amin's, all published in the 1980s, but find that in each case there is doubt
concerning the magnitudes obtained. They conclude: 'that to date there has
been no satisfactory calculation of unequal exchange'. The reason why this is so
has been hinted at but not developed in theoretical terms. The Appendix
provides a brief exposition of the theory of unequal exchange for the reader
who is interested. That exposition serves to expose the insubstantial basis for
the doctrine.

Nevertheless, the idea of unequal exchange clearly has certain attractions, in
that the source of underdevelopment is located externally to the country in
question, within the operation of the global capitalist system. In this sense, the
blame for the poverty of the poor lies largely with the rich. Furthermore, the
doctrine has a clear and ready extension from the international setting to the

relationships of regions within nations. Peripheral regions and smaller cities, it is held, suffer a net loss of resources, while the centre (usually the capital) is regarded as being a net beneficiary.

## The supply-side counter-revolution

In the mainstream of economic debate, the most formidable challenge to Keynesian orthodoxy has come from monetarist thinkers rather than those of Marxist persuasion. This challenge arose in response to the perceived failure of Keynesian economics to account for two manifest problems and to provide the requisite answers.

The first of these problems was the difficulty which governments experienced in controlling the level of demand. In the British case, Pollard (1982, p. 51) records nine crisis occasions in the period from 1947 to 1973–6, all triggered by balance of payments problems. The emergency measures which it was necessary to take were deflationary, and came to be known as the 'stop' phase of 'stop-go'. Part of the problem was that during the 'go' phase, the volume of imports rose much faster than exports, revealing a weakness in the 'mechanical' version of Keynes' thinking, in that the economy cannot be treated as if it were closed. The actual experience was one of sharp oscillations in output and, to a lesser extent, unemployment, about a long-term trend characterized by slow growth of output. The economic oscillations and the accompanying sharp changes in policy inevitably called into question the possibility of fine-tuning an economy in the manner visualized in Keynesian thought.

Although 'stop-go' was a more serious problem in Britain than in most other developed countries, similar difficulties had been experienced elsewhere. However, doubts concerning the ability of governments to fine-tune their economies might have remained submerged if a reasonable tempo of overall growth had been maintained. But with the 1970s came also 'stagflation', the combination of high (and rising) levels of unemployment and high (and accelerating) rates of inflation. In varying degree, this phenomenon was experienced throughout the developed world, and especially following the oil-price rise of 1973–4. But the warning signs had been observed almost a decade earlier by a handful of scholars, as, for example, Thompson (1965): 'the nature of the unemployment problem seems to be changing. For some years now, Keynesians have been haunted by the paradox of rising prices (too much spending?) in periods of unemployment (too little spending?) (1965, p. 206). In a nutshell, additional expenditure was not being translated into higher domestic output and larger numbers in employment, as the current orthodoxy would have had one suppose to be the case. Instead, in many countries, but most notably in the United States and Britain (Solomon, 1984), it was being translated into upward pressure on prices, a process which was given great impetus by the 1973–4 rise in oil prices. By the middle of the 1970s, it was widely accepted that something was seriously amiss with the current economic orthodoxy.

One reaction to this situation has been to explore the Keynesian doctrine to identify weaknesses which could be rectified, or missing ingredients which should be included. One of the more notable examples of this endeavour is Meade's (1982) exploration of wage policies. The central problem addressed by Meade is the question: How can wages be determined in a manner that will permit steady growth at full employment without excessive inflation? More generally, as Kaufman (1988) points out, Keynesian scholars have taken the view that macro-theory is in trouble because the micro-theoretical foundation is itself erroneous.

The second reaction has been to argue that Keynesian doctrine is fundamentally flawed because it is not solidly constructed on existing micro-theory, which is regarded as sound, and because it emphazies the wrong elements of the macro-economic system. M. Friedman (1956, 1985) has been a leading exponent of this view. His central contribution has been the argument that the main determinant of nominal aggregate demand is the supply of money and not the Keynesian categories of expenditure – e.g., investment and government expenditure – which governments have used in their efforts to control the economy. According to Friedman's analysis, rapid inflation implies an excess of money in relation to the available supply of goods and services. Therefore, to control inflation one must limit the availability of money.

It is, of course, an open question whether one should approach the problem of prices through the supply of money or through the level of demand for money. However, the main point is that monetarists emphasize the role of money supply in the economy, whereas Keynesians generally regarded money as unimportant. Similarly, monetarists are doubtful concerning the utility of fiscal measures to control aggregate demand and prefer to use the rate of interest as the major weapon for this purpose (Feinstein, 1983).

Friedman's second major insight concerned the wage-bargaining process. The famous Phillips curve, published in 1958, apparently showed that a trade-off is possible between the level of unemployment and the rate of inflation. However, to derive this conclusion from the historical evidence marshalled by Phillips, one must assume that workers bargain for nominal wages. Under inflationary conditions, workers will learn to bargain for real wage increases, and will learn to do so with reference to expected future inflation in addition to the inflation of the recent past. If there is a 'natural' rate of unemployment, any attempt by the authorities to reduce unemployment below this level will only serve to increase wages and prices. Consequently, it is vain to attempt to buy lower unemployment with higher inflation, since all that one gets is yet higher inflation, and possibly also yet more inflation. Friedman had drawn attention to this problem in 1967. Events quickly seemed to confirm his predictions, with the result that monetarist doctrine was given considerable credibility (however, the experience of the 1980s then showed that there are considerable difficulties in positing a *stable* 'natural' rate of unemployment, or non-accelerating inflation rate of unemployment as it came to be called – *Economist*, 13 August 1988, pp. 15–16).

It was the emphasis on money supply which gave the counter-revolution its name – monetarism – although Friedman himself prefers the old-fashioned term 'the quantity theory of money'. Associated with the reinstatement of money as a variable which governments should seek to control has been the emergence of a wider concern with supply-side issues. Conceived in narrow terms, this concern is limited to one issue in addition to the supply of money, namely the role of direct taxes. If tax rates are raised, the immediate effect is to reduce the money available to individuals to spend. Within the Keynesian tradition, it is assumed that taxpayers, striving to maintain their post-tax income, will work longer hours at their main job or will take a second employment. On this interpretation, higher taxes lead to more personal effort and a bigger national income. Supply-side thinkers point to an alternative response. A higher tax rate implies a lower marginal return to effort. Unless preference functions shift, workers will choose to take more leisure and to work fewer hours. Supply-side economists point to the long-run disincentive effects of high marginal tax rates and therefore advocate their reduction as a way to increase output and employment. These conflicting interpretations have been thoughtfully discussed by Ture (1982) in a non-polemical way. Meantime, in both the United States and in Britain, considerable cuts have been made in the higher rates of personal tax (from levels in excess of 80 per cent in some cases), in the belief that this will increase the incentive to work. As justification, both the Reagan and Thatcher administrations argued that the increase in personal effort would generate more activity in the economy, thereby raising total employment and generating additional government revenue to replace that forgone by the lowering of taxes. In the event, it does appear that tax revenues actually paid by the wealthier earners have increased in comparison with what they would have been at the previous tax levels. In the context of the 1988 presidential election in the United States, *Economist* (19 March 1988, p. 30) observed: 'All [candidates] would judge that top tax rates much above 40 per cent would bring lower employment rather than more revenue.'

To emphasize money supply and taxation is to emphasize the doctrinaire aspects of what is loosely known as monetarism. In fact, supply-side issues run much wider (Drucker, 1981; Laidler, 1981). The underlying concern is with the long-run productive capacity of the economy, allied to a deep scepticism about the ability of 'economist kings' to make the right decisions to fine-tune an economy. The fundamental stance, therefore, is the belief that governments should use monetary means to establish a predictable environment in which individuals and firms can make sensible judgements about the future and act accordingly, subject to constraints on that behaviour which are deemed to be in the public interest. This in turn places the focus on the markets for the factors of production – for land, labour and capital. The question to be asked is: Where are there bottlenecks in supply, and how might they be eased or removed? Viewed in these terms, the debate between Keynesian economists, monetarists and supply-side advocates reduces to the problem of what emphasis to give to the various economic elements at the macro-economic level, and the place to be accorded micro-economic issues in the macro-economic context. 'Keynesianism, monetarism, and supply-side economics are not conflicting theories of how the economy works, but limited perspectives stressing different factors' (Feulner, 1983, p. 263).

So what are the implications of this debate which are relevant for regional economic analysis? The supply-side stance portrayed in terms of a stable monetary policy and the incentive effects of reduced taxation does not itself translate into regional implications, except through the operation of regionally differentiated personal and property taxes. Nor, at face value, does the wider interest in the supply of the factors of production, except as a latter-day version of neo-classical doctrine, i.e., that supply bottlenecks should be eased or removed wherever possible. There is, however, this difference. Neo-classical thought assumed that factors of production are fully mobile and that rigidities are the exception. Supply-side economists make no such assumption. Indeed, they emphasize that it is precisely the fact of rigidities and imperfections in the factor markets that inhibits a faster rate of overall growth. Furthermore, if one accepts this notion – not necessarily as proven but at least as worthy of consideration – then a whole new dimension of analysis is opened up. Labour, in particular, is not 'sold' on a national market but within local job catchment areas, with limited short-term possibilities for migration. If there are inefficiencies in the labour market, these arise from inefficiencies in the myriad of local labour markets. The cause of such inefficiencies may lie as much in regional as in national conditions; by the same token, the remedy may lie in either the regional or the national domain. The same observation may be made about the operation of capital and land markets, especially in federal nations such as the United States, but by no means exclusively. All of this suggests an important new dimension to thinking about regional growth processes.

But perhaps the most important insight derives from the following. Keynesian doctrine emphasizes the short run, in which coefficients are stable and technology is constant. Keynes himself believed that he had disproved Say's Law, and Keynesian thinking tended to downplay the role of supply. Drucker (1981) points out that, as a consequence, Keynesians assumed that productivity would look after itself if aggregate demand were managed appropriately. For the United States and the United Kingdom, both of which had been managed *as if* they were macro-economies (i.e., closed), the productivity trends up to the early 1980s, in contrast with countries such as Germany and Japan, had been unfavourable. For both the United States and the United Kingdom the competitive position had been deteriorating. Given the growing interdependence of the world's manufacturing countries, these adverse trends could not be accepted with equanimity. To improve the situation, however, implied explicitly recognizing that supply-side matters could not be left to look after themselves.

From a regional perspective, the operation of factor markets and supply availability could have a bearing on any or all of the following:

(1)  The relative efficiency of production in a given product line, and hence competitiveness with other regions.
(2)  The facility or otherwise with which a regional economy adjusts to changes in comparative advantage, abandoning less promising sectors (products) and taking up others with better prospects.
(3)  The facility with which a region spawns entirely new enterprises which offer new products and create their own markets.

All three offer the possibility that, at the regional level, supply can create its own demand – by quality and price (equals efficiency), or by virtue of novelty. Technological leads and lags become crucial, as does the role of entrepreneurial talent – Schumpeter's fourth factor of production.

## Conclusion

In a single chapter, it is impossible to do full justice to the richness and variety of economic doctrine. The purpose, however, is not to recount doctrine for its own sake, but to provide a framework within which to approach the problems posed by regional growth, or the lack thereof, and the ways in which growth can be stimulated or controlled. The literature which deals with these problems derives most of its general theory from one or other of the five major schools of thought which have been reviewed in this chapter. Because these schools differ in the emphasis which they place on relevant variables, the interpretation of regional economic prosperity and distress shows considerable variety. If one can recognize the source of inspiration, one is better placed to assess the utility of the regional analyses which are on offer, and the circumstances under which they may be more or less useful. In the succeeding chapters, we will examine in more detail how the main schools of economic thought translate into the spatial domain and lead to conclusions for policy.

# CHAPTER 4

# *Neo-classical and Keynesian theories of regional growth*

If you pick up a text on regional economies written a few years ago, you will find it dominated by short-run macro-economic demand models and their derivatives.

(Richardson, 1978, p. 82)

Two schools of thought have, until relatively recently, dominated work on regional economics and regional growth – the neo-classical and the Keynesian. Therefore, it is appropriate to commence our discussion of regional growth with these two bodies of doctrine. The approach to be adopted is as follows. Each school of economic thought will be examined for the implications it has for the way in which regional growth is visualized to occur, and the need for government intervention, and the means to be employed. As will become abundantly clear, the neo-classical and Keynesian schools of thought lead in quite different directions so far as appropriate policy conclusions are concerned. These differences arise from the divergent assumptions which form the basis of theory. Consequently, an appeal to empirical evidence cannot be expected to determine which is 'right' or 'wrong'. However, the experience of recent decades does throw some light on the matter. This material will be considered in Chapter 5, which shows that neither approach is fully satisfactory in explaining what has actually been happening in the industrialized countries. This is a partial explanation for the dissatisfaction which has been widely articulated about the state of regional growth theory and the nature of policy, and paves the way for the consideration of other schools of thought in subsequent chapters.

Economic theories can be divided into two classes – macro- and micro-theory. Macro-theory is concerned with national aggregates – GNP, unemployment, etc. – and their interrelationships. Micro-theory, in contrast, deals with the behaviour of individuals and firms. Now a region lies part way between these two levels of analysis, and there is no true economic meso-theory. We are therefore faced with the uncomfortable fact that, in tackling regional economies, we may have to make unsatisfactory adaptations from a particular body of economic doctrine. Neo-classical theory is micro-economic in character, whereas the Keynesian system deals with macro-economic matters. The former

**Table 4.1**   Macro-economic policy measures with regional implications

| Category | Regional effect |
|---|---|
| Fiscal: | |
|    Automatic stabilizers | Progressive taxes and income support measures (especially unemployment benefit) |
|    Discretionary | Regional variation in taxes and central government expenditure (including infrastructure and procurement) |
| Monetary | Geographical variation in interest rates and credit control |
| Import controls | Protect specific industries, which may be localized |
| Export controls | Assist specific industries, which may be localized |
| Currency exchange rate | Affects the competitiveness of domestic production and exports relative to imports |
| Public investment | Differential regional impact |

*Source*: adapted from Armstrong and Taylor, 1978 and 1985

body of doctrine makes certain assumptions regarding macro-economic circumstances, while the latter similarly assumes the micro-economic system to have specified characteristics. Some aspects of this parallelism are explored in Tables 4.1 and 4.2, which set out regional aspects of macro- and micro-policy measures.

At this preliminary level, we will finally note that, despite the Keynesian revolution, neo-classical thought has been remarkably tenacious in the field of regional economics (McCombie, 1988a; Richardson, 1978). As a bald generalization, it has often been held that neo-classical theory provides a normative view of how regional economies should operate, whereas Keynesian doctrine provides a better account of what actually does occur. To the extent that normative is compared with positive theory, we are dealing with non-commensurable ideas. By the same token, however, Keynesian thinking has seemed to yield much more practical and applicable theory, providing the basis for the comment with which this chapter starts – i.e., that Keynesian thought has been dominant.

## Neo-classical regional growth

### Theory

Two components of regional growth are visualized – long-term growth on the one hand, and short- to medium-term growth associated with a transition from initial disequilibrium to an equilibrium state. The latter process involves the interregional migration of labour and capital, so that the factors of production may obtain identical returns in all locations. As a consequence, some regions

**Table 4.2**  Micro-economic regional policy measures

| | |
|---|---|
| *Policies to reallocate labour* | |
| *In situ* reallocation | Occupational retraining |
| | Education policies |
| | Journey-to-work subsidies |
| Spatial reallocation | Migration policies |
| | Assistance with housing |
| | Job information |
| | Improvement in efficiency of labour markets |
| *Policies to reallocate capital* | |
| Taxes and subsidies: inputs | Assistance with capital investment |
| | Wage subsidies |
| | Operational subsidies, e.g., fuel, transport |
| Taxes and subsidies: outputs | Export rebates |
| | Price subsidies |
| Taxes and subsidies: technology | Research and development |
| | Innovation |
| Improved efficiency of capital | Loan guarantees |
| markets | Export credit guarantees |
| | Venture capital |
| Adminstrative controls | Controls on location of investment |
| | Planning control administration |
| | Reduced bureaucratic/administrative control |

*Source*: adapted from Armstrong and Taylor, 1978 and 1985

will grow more rapidly than the average, while others will grow more slowly (or even experience negative growth). This spatial reallocation process is conceived in terms which are consistent with the factor proportions theorem of international trade. With the passage of time, the differential rates of growth occasioned by interregional reallocation will converge to zero, and all regions will experience only the long-run growth, which is determined by the rate of population growth and the rate of technical progress. Net reproduction and technical change are assumed to be exogenously determined, i.e., to be factors which are taken as given and for which no explanation is offered.

Neo-classical regional growth theory has been described by many authors, who may be consulted for further details (e.g., Armstrong and Taylor, 1985; McCombie, 1988a; Richardson, 1973, 1978 and 1984). Although the assumptions are derived from neo-classical economic thought, the introduction of space implies that there are costs associated with the reallocation of the factors of production, with the movement of goods and with the transmission of information. These costs are serious violations of some of the assumptions on which neo-classical thought is based. Indeed, as Richardson noted: 'if we were to adopt neo-classical models in their pure unadulterated form there would be no such field as regional economics' (1973, p. 23). Consequently, the strategy in constructing a model of regional growth is to relax as few of the basic assumptions of neo-classical thought as is possible, and then to assume that the relaxations do not imply an unacceptable level of inconsistency in the resulting theoretical structure.

For the purpose of exposition, it is convenient to begin with the problem of long-run growth, and then subsequently to incorporate the interregional reallocation process as the system moves to equilibrium growth. Let us assume that production functions are identical in all regions and that there are constant returns to scale, i.e., no economies of scale exist in the system. Let us further assume that, in the long run, labour resources are fully employed and that returns to labour and capital are identical in all regions (and in all sectors if there are more than one). The long-term rate at which a single region (i) grows can be written as:

$$y_i = a_i k_i + (1 - a_i) l_i + t_i \tag{1}$$

where the subscript i denotes the particular region, and the following stand for the growth rates of:

$$y = \text{output}$$
$$k = \text{capital}$$
$$l = \text{labour}$$
$$t = \text{technical progress}$$

The coefficient, a, is written as a decimal fraction, e.g., 0.6, so that the terms $a_i$ and $(1 - a_i)$ express the ratio of capital to labour.

The amount of capital and of labour in a region depends on two things: the locally generated supply and the supply contributed by factor migration. Consequently, we may write:

$$k_i = \frac{s_i}{v_i} \pm \sum k_{ji} \tag{2}$$

where

$$s = \text{saving/income ratio}$$
$$v = \text{capital/output ratio}$$
$$k_{ji} = \text{annual net capital flow from region j to}$$
$$\text{region i as a proportion of region i's}$$
$$\text{capital stock}$$

and

$$l_i = n_i \pm \sum m_{ji} \tag{3}$$

where

$n$    = rate of increase in the indigenous labour
supply

$m_{ji}$ = annual net migration from region $j$ to
region $i$ as a proportion of region $i$'s
labour supply

The migration rates incorporated in equations (2) and (3) will be determined by the regional differences in the rates of return to capital and labour. These differences may represent the initial disequilibrium position, in which case they will disappear with the passage of time. If in fact the differences persist, then there is a long-term divergence from the basic assumptions of neo-classical theory. We may express the rates of migration as

$$k_{ji} = f(r_i - r_j) \tag{4}$$

and

$$m_{ji} = f(w_i - w_j) \tag{5}$$

where

$r$    = rate of return to capital
$w$   = wage rate

Note that, as presented thus far, it is implicitly assumed that capital and labour can move between regions at zero cost. In this sense, the model retains the neo-classical assumption that economic activity occurs with no friction of distance. Such an assumption is clearly unrealistic, especially for growth in a system of regions. However, it is analogous to the idea incorporated in the Ricardian and factor proportions approaches to international trade, in which it is supposed that, whereas the factors of production are immobile between countries, goods may move without cost between countries, and the factors of production may similarly move between sectors.

Equations (1) to (5) can be modified to bring the model nearer to reality, notably by incorporating transfer costs for the factors of production which move between regions and by expanding the production system from one sector to several. A multi-sector version must include explicit assumptions regarding the intra-regional cost of transferring resources from one sector to another. However, the more 'realistic' that the model is made, the further will it diverge from the basic assumptions of neo-classical economic thinking. Uncomfortable inconsistencies are introduced into the model, and the question has to be asked whether an alternative starting point would not be more suitable.

The main interest of the neo-classical model lies in the predictions which can be derived regarding the reallocation of resources as a regional system adjusts to an initial disequilibrium situation. On the assumption that production functions are identical in all regions and exhibit no returns to scale, initial regional differences in the rates of return to capital and of wage rates for labour will depend on variations in the capital/labour ratio. It is then axiomatic that high returns to capital imply low wage rates, and vice versa. By definition, therefore, the reallocation process must involve the migration of capital in the *opposite* direction to the movement of labour. This conclusion is axiomatically true for a regional model which incorporates only one production sector. The same conclusion is generally accepted for models with two or more sectors (however, see McCombie, 1988a, p. 276). Consequently it is axiomatic that incomes (and unemployment rates) as well as returns to capital will converge to the same values and thereafter growth rates will be determined by population growth and technical change alone.

A final comment is in order. Neo-classical regional growth theory is generally described as being a supply-side model (Armstrong and Taylor, 1985; Richardson, 1969; Thirlwall, 1980). This is clearly true, in the sense that long-run growth is assumed to be determined by the supply of people (population growth) and of technical innovations. These are treated as autonomous variables, which implies that any questions there may be regarding these supply-side variables are assumed away. Furthermore, since the factors of production (capital and labour) are assumed to be homogeneous and freely mobile, any supply-side problems within the system are ignored. As there are in fact numerous supply-side matters which are not addressed in the neo-classical model, it is unfortunate that it has acquired the reputation for being supply-side in its formulation.

## Policy conclusions

If an approximately 'pure' version of neo-classical thought is accepted, it is axiomatic that the allocation of resources should be left to market mechanisms. These are perceived to be simultaneously the most efficient means for ensuring that resources are put to their best uses and also sufficient for achieving this end. Consequently, pure neo-classical thought holds that an initial disequilibrium will disappear of its own accord and that therefore government should stand to one side, while 'the market' sorts things out. Intervention by government is perceived to be unnecessary, and were intervention to occur, the end result would be sub-optimal, because administrative action, it is held, must be less efficient than market forces.

Unfortunately, the real world does not always behave in the manner predicted by this theory (see Chapter 5), and some modification to the austere rigour of neo-classical thought is common. The fact that disequilibrium persists for long periods of time may be attributed to a failure of the relevant markets to operate with full efficiency. Comparing capital and labour, it is widely perceived that capital can adjust more readily than labour. Over a period as short as five or ten years, the retirement of obsolete or redundant capital (in the form of plant and buildings), plus the pattern of new investment, can have a considerable impact on the geography of capital. In contrast, numerous barriers may limit the ability of people to move from one region to another – social ties, housing, schooling of children, etc. Therefore, market failure is perceived to be largely the failure of labour to make the requisite adjustments. Although, within neo-classical thought, it is illogical to assume that one factor of production is less mobile than another, the pragmatic recognition of realities suggests that there are in fact considerable differences between the two factors.

If the spotlight focuses on the inadequate adjustments in the labour market, the interpretation of the market failure is highly conditioned by the basic assumption that the factors of production are homogeneous. Consequently, a labour force in a region, faced with a dearth of jobs locally, has, in the neo-classical perception, two options. The first is to migrate elsewhere, to regions where jobs are more plentiful – in Mr Norman Tebbitt's notorious words, 'Get on yer bike'. The alternative is to accept lower wages, the effect of which would be to expand total employment. If, taken together, these two responses prove to be inadequate, then government should consider the implementation of policies designed to achieve either:

(a)　more rapid interregional migration; or
(b)　more drastic wage reductions in the regions of high unemployment.

Of these options, it is the first which has generally been preferred by governments, being politically more acceptable and easier to implement.

Common observation shows that neither capital nor the labour force are in fact homogeneous. However, the neo-classical tradition ignores this manifest reality; by assumption, these differences do not exist. Thus, in their concluding comment on neo-classical growth models, Armstrong and Taylor note: 'The lack of realism of many of the fundamental assumptions of neo-classical models of regional growth has induced some researchers to reject the neo-classical approach and to search for entirely different explanations of why regions grow at different rates' (1985, p. 64). They emphasize that in a two-sector (or n-sector) version of the neo-classical regional growth model, each region must engage in trade – both imports and exports. The fact of exports implies the existence of demand which is external to the region in question, a matter which is emphasized in Keynesian thinking.

## Keynesian regional growth

Keynes himself gave almost no attention to regional problems, his only explicit contribution being articles on 'How to avoid a slump' in *The Times* in 1937. However, a distinctively Keynesian view of regional development became prominent in the post-war period, drawing upon Keynesian thinking and the work of other scholars, and emphasizing the disequilibrium nature of regional growth processes (Armstrong and Taylor, 1978 and 1985; Brown, 1972; Dixon and Thirlwall, 1975; Kaldor, 1970, 1972, 1975 and 1981). The Keynesian foundations can be clearly recognized in the role assigned to demand and to multipliers. Together, these provide the foundations for analysis. The superstructure is built from economies of scale and Verdoorn's effect; while neither concept is specifically Keynesian in origin, both are consistent with Keynesian thought, in that both lead to long-term disequilibrium processes. What we will call the Keynesian regional model is one that stands clearly in that tradition but is not something that he himself developed. As it evolved in the post-war years, it was quickly seen as providing a general theoretical framework uniting several strands of thought which developed quite independently of Keynes and his followers, and indeed before the regional model had been fully worked out. This fact contributed to the popularity of the model.

## Antecedents

The evolution of the Canadian economy was described by Innis (1930 and 1936) in terms of 'staple' products, such as furs, timber and minerals, the development and export of which provided the basis for local and national development. His ideas were taken up by North (1955), who elaborated a growth process propelled by exports. Once some initial export-oriented development has occurred, the existence of transport and infrastructure may make other export enterprises viable. In any case, the population engaged in export activities will require food, clothing and other necessities; initially, these will be imported but local production may soon replace the imports. Consequently, the impulse provided by the initial export industry may trigger an upward spiral of development.

A convergent strand of thought emerged among town planners and urban analysts during the 1930s and was codified by Andrews (1953-6) and Pfouts (1960) in the concept of basic and non-basic employment. For the management of an urban area, it is important to predict the future size of the population. If one disaggregates total employment, one may postulate that the basic sector comprises those activities which export goods and services and thereby generate income. The non-basic employment provides for the needs of the basic sector and also of the population generally. If the basic/non-basic ratio can be established, it will be possible to assess the local impact of predicted future changes in basic employment.

Another convergent strand of thought was provided by the work of Perroux (1950, 1955; see Higgins and Savoie (eds), 1988) in the early 1950s, followed later that decade by Hirschman (1958) and Myrdal (1957). The corpus of ideas attributable to these and other authors generally goes by the name 'cumulative causation', a development process which is often, but mistakenly, identified only with Myrdal (Gaile, 1980). Perroux himself was much influenced by Schumpeter's emphasis on innovation as the basis for development, and the associated idea of 'propulsive industry'. However, that aspect of cumulative causation was given less prominence in the 1960s and later. A common presentation of the cumulative growth process is that shown in Figure 4.1, in which the initial stimulus is provided by the location of a sizeable employer, exporting goods (or services) from the region. Once the new firm is established, linkage and multiplier processes are supposed to operate, stimulating other firms in the area, and driving regional (or national) output to higher levels. In addition, the

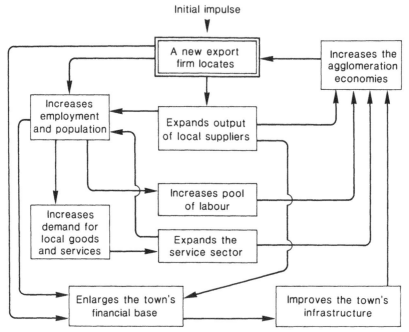

**Figure 4.1**  Cumulative regional growth

idea of cumulative growth places much emphasis on economies of scale, shown in Figure 4.1 as 'agglomeration economies' (cf. Figure 3.4).

The concept of scale economies has been familiar since Adam Smith used the manufacture of pins to illustrate the principle but, as we have seen, it fell out of favour with the ascendancy of neo-classical thinking. The idea was given some prominence in the work of Chamberlin and E. H. G. Robinson in the 1930s, and had some impact on the literature concerned with location, trade and development (Ohlin, 1933; Weber, 1929; Young, 1928). However, the redis-covery of scale economies in a regional context only began in earnest with the 1940 Report of the Royal Commission on problems of regional development in Britain, chaired by Sir Montague Barlow. Then, shortly after the Second World War, several important publications laid the groundwork for the recent interest in the role of scale economies in regional development (Chisholm, 1966; Clark, 1967; Isard, 1956; Robinson (ed.), 1960; and Youngson, 1967).

The last converging strand of analysis derived from geographical work on transport costs as a factor in location. Noting that the importance of industries oriented to raw materials was declining, and that access to markets was becoming more important in both absolute and relative terms, Harris (1954) argued that generalized accessibility measures – or market potential – provided an important tool for examining trends in regional development. In this seminal paper, he argued that because market access was an increasingly important factor, cumulative concentration was occurring:

manufacturing has developed partly in areas or regions of largest markets and in turn the size of these markets has been augmented and other favourable conditions have been developed by the very growth of this industry.

(Harris, 1954, p. 315)

Subsequently, numerous workers took up the idea of market potential as a tool with which to analyse regional development, in the expectation that growth will occur in a cumulative fashion in those areas which have the greatest accessibility (e.g., Clark, 1966; Clark *et al.*, 1969; Keeble *et al.*, 1982).

These various strands of thought all point toward regional growth as a disequilibrium process, contrary to the propositions of neo-classical thought. They also focus on the role of exports from a region as the motor for regional growth. Exports, in turn, may be viewed as dependent on the role of demand. Clearly, the Keynesian emphasis on the role of demand, allied with the assumption that economic systems do not necessarily move to an equilibrium at full employment without intervention, provides an attractive way in which to try to systematize these various ideas. It was Kaldor in particular who formulated them in a series of influential papers, published mostly in the 1970s (Kaldor, 1970, 1972, 1975 and 1981). One of his major contributions was to focus attention on the role of export multipliers. His ideas were formalized by Dixon and Thirlwall (1975) and became part of a widely accepted view of regional development (e.g. Armstrong and Taylor, 1985; Brown and Burrows, 1977; Moore and Rhodes, 1982; Thirlwall, 1980).

## Theory: multipliers

For the management of the national economy, it is essential to know what the overall impact of a given change in demand will be. Therefore, central to the practical implementation of Keynesian demand management is knowledge of the relevant multiplier effects. Translated into the regional domain, the driving force for growth can be visualized as changes in income arising from either of two causes, plus the effects which arise from the associated multiplier. The two primary sources of income change are visualized to be export earnings and government expenditure. Although both are clearly important, greater emphasis has been given to the export multiplier. One of the strongest statements on record in favour of this approach was written as recently as 1980:

For a region in which capital and labour are highly mobile in and out, growth must be demand-determined. If the demand for a region's output is strong, labour and capital will migrate to the region to the benefit of that region and to the detriment of others. Supply adjusts to demand. We cannot return to the pre-Keynesian view that demand adjusts to supply. If we could, the solution to any region's lagging growth rate would be for it to save more and breed more! Production function studies of interregional growth performance, which approach growth from the supply side, have their uses, but the question still has to be answered: why does the rate of growth of labour, capital and total factor productivity differ between regions? The major explanation must lie in differences in the strength of demand for regions' products. The only true supply constraint on growth is land-based resources, but economic activity in most regions in mature economies is not land-based.

(Thirlwall, 1980, p. 420)

The income which is generated by such exports becomes, at the next round, expenditure, which in turn becomes income for yet other people.

As a first approximation, a region can be said to have two economic sectors – the export (basic) sector and the non-export (non-basic) sector. If we assume that the magnitude of the non-export sector is determined by the income (employment) in the export sector, total regional activity will be some function of the scale of activity in the export sector. Precisely what the relationship will be depends on the amount of income that is saved and the proportion of expenditure that is directed to imports. In highly simplified terms, we may express the export multiplier as:

$$y_i = \frac{1}{m} x_i$$

where

$$
\begin{aligned}
y &= \text{income} \\
x &= \text{exports} \\
m &= \text{the propensity to import} \\
\frac{1}{m} &= \text{the export multiplier}
\end{aligned}
$$

i denotes a particular region.

For example, if the propensity to import is 0.5, income will be twice the value of exports; were it to be 0.25, a given value of exports would result in an income four times as large.

Although conceptually clear and elegant, the actual use of multipliers is fraught with difficulties. The distinction between export and non-export activities is far from clear cut, and it is difficult to estimate the magnitude of second-round effects, as the imports into a region generate income elsewhere which results in additional demand for exports from the region in question. In any case, multipliers operate within constraints, which may be constraints on the supply of the factors of production or, as Thirlwall (1980) has argued, balance-of-payments constraints. However, the view has been put, by Thirlwall among others, that supply constraints are not important in a regional system in which labour and capital are free to migrate.

Issues of this kind may be set to one side for the moment. The unqualified export multiplier doctrine treats the demand for a region's exports as being exogenously determined (demand is said to be autonomous), with the consequence that: 'According to Kaldor, regional growth is fundamentally determined by the growth of demand for exports' (Dixon and Thirlwall, 1975, p. 203). Five years later, Thirlwall (1980) expressed the relationship in almost identical terms – that the regional growth rate in balance-of-payments equilibrium approximates to the growth of regional exports divided by the regional income elasticity of demand for imports. Hence: 'so long as some roughly known value for it [the multiplier] can be relied upon, growth of employment in a region can be said to depend solely upon growth of basic employment in it [the region]' (Brown and Burrows, 1977, p. 35).

This highly stylized view of regional growth ignores the qualification which Kaldor himself made in his 1970 paper – that the share of the 'autonomous' demand which a region captures depends on its costs of production relative to costs of production elsewhere. Production costs can be viewed as a resultant of productivity and of wages, both of which are, at least in part, influenced by action within each region. More important, though, has been the widespread assumption that the only link between imports and exports is through the income effect arising from changes in export earnings. In addition, of course, imports are necessary to provide the materials from which exports are manufactured (input-output relationships) and, more important, the volume of both imports and exports is intimately affected by the cost, or more generally, efficiency, of production. If a region captures a large share of 'autonomous' demand, it should also be able to compete effectively against imports over some range of goods (see Chisholm, 1985a; Rowthorn and Wells, 1987).

One other qualification is in order at this juncture. External demand, which, on a first approximation, can be treated as being autonomous, is clearly of immense importance for a small economy which is highly dependent on a limited range of goods, such as the Mesabi Iron Range in northern Minnesota (USA). Reduced demand for its iron ore has sharply curtailed the activity and prosperity of this region. In contrast, California has a population approaching 30 million, larger than many nations, and possesses a diversified manufacturing economy, much of which operates at technological frontiers. Export multipliers are of much less significance in the latter than in the former case (cf. Figure 1.1).

Despite the caveats, it is self-evident that multiplier processes – whether general income multipliers or export multipliers – must exist and must be important. They provide a powerful explanation for the regional problems which arise as staple industries decline, triggering negative multiplier effects (e.g., New England textiles, and the coal and steel industries of the Ruhr). Conversely, the success of 'high tech' firms in California and southern England generates further pressure for growth. As the effects of an initial impulse work through the system, multipliers (positive or negative) will tend to accentuate the initial difference, exacerbating the stagnation and decline of some regions, and the pressure for growth elsewhere. These effects lend powerful support for the disequilibrium view of regional development.

## *Theory: interregional models*

Multiplier analysis is commonly undertaken in terms of two regions only – the region in question and the whole of the rest of the country (even world). Interregional models extend this approach, in the form of an n-region model for an entire nation, usually with one region to represent the rest of the world. In such a model, the interrelations of all the regions and sectors are specified in some degree (Richardson, 1985). Conceptually, we may distinguish between input-output models and econometric models, though the distinction is in practice somewhat blurred (Rietveld, 1982).

An interregional input-output model is, in essence, a matrix of coefficients which describes the links between industries within and among several regions. Such a model allows one to predict the change in industrial and regional activity for a given change in final demand (or GNP). At the simplest, it is assumed that there are no supply constraints, so that output can expand indefinitely. In practice, output limits will be set by the availability of labour and capital, and possibly by other input constraints. Therefore, an input-output model is of limited utility unless it incorporates factor supply constraints. It is but a short step to add a migration component for labour and capital. Other variables can be added, as, for example, the effect of unemployment rates upon wage rates. The more fully an input-output model is specified, the more difficult does it become to distinguish it from a full econometric interregional model.

It was Leontief in particular who pioneered input-output analysis, including the regional application of the idea; in the latter context, he worked closely with Isard (Isard, 1951; Isard *et al.*, 1966; Leontief, 1951; Leontief *et al.*, 1953). Apart from the formidable data problems, even for reduced versions of interregional models, the main points of interest in the present context are as follows. Some early writers grappled with supply-side issues in terms other than just the constraints on output imposed by supply limits (e.g. Isard, 1951), or argued that the implications of market organization should be taken into account (Chinitz, 1961 and 1966). However, such considerations were submerged by the demand-driven nature of practically all the models that have been developed. Glickman, writing in 1982 (p. 100), remarked: 'Almost all regional modeling to date has been demand-oriented.' A similar assessment had been offered a few years earlier by Engle:

> The growth and development of regional economies can be broadly char-acterized as depending on three external forces: the demand for the products produced, the supply or migration of labor to the region, and the supply or migration of capital to the region. Product demand has tradi-tionally and econometrically been considered the primary feature,

although factor migration is increasingly given a major role in the explanation for some of the regional patterns observed in the U.S.

(Engle, 1979, pp. 157–8)

In addition to being demand-driven, most interregional models are short-run in character, because of the difficulty of building in changes in the input-output and econometric relationships as the economy evolves. In this sense, these models – along with multiplier analysis – can be regarded as fitting the tradition of the 'mechanical' Keynes. This generally fixed-coefficient approach can alternatively be treated as a short-run approach, in contrast to the longer-run analysis which is generally characterized as neo-classical (Foster and Mulley, 1988).

## *Theory: the Verdoorn effect*

In an advanced economy, it is reasonable to assume that, as capital invest-. ment occurs, the resulting production facility will embody the latest technology. As a result, recently installed plant will allow production to occur at unit costs which are lower than can be obtained with older equipment. Alternatively, higher quality output is possible at a given cost. It follows that, if the average age of the capital stock in one region is lower than in other regions, it will be able to offer cheaper and/or better commodities and services than are available elsewhere. This effect applies to the capital stock of individual firms and also to the capital which provides common services – transport infrastructure, utilities and education/training facilities, etc. To the extent that regionally supplied inputs are more economically available in one region than another, firms may find it advantageous to expand in or migrate to the more favoured region.

At the international level, there are marked differences in the rate at which capital equipment is replaced – about 2 per cent *per annum* in the Soviet Union, 5 per cent *per annum* in the United States and 10 per cent *per annum* in Japan (*Economist*, 18 November 1989, p. 25). Estimation of replacement rates in the different regions of a country is much harder than estimation of national aggregate rates and has been attempted only rarely. However, figures for Canada (Anderson and Rigby, 1989) indicate that whereas in Alberta the 1981 average age was only 8.53 years, in British Columbia the figure was 11.13 years; all the other provinces and regions lay within this range. If the productivity of capital was growing at 2 per cent per annum, an age difference between the two provinces of 2.6 years would have given a difference of over 5 per cent in capital efficiency, a difference sufficient to confer a considerable advantage on Alberta.

The Verdoorn effect, therefore, provides a mechanism whereby a region which grows faster than other regions will obtain a productivity or quality advantage which will probably mean that it will continue to grow more rapidly – through the growth of existing firms and by the in-migration of firms. Thereby, it reinforces the disequilibrium effects which arise from the operation of multipliers.

## Theory: scale economies

The concept of scale economies is well known and needs little rehearsal. Internal economies may arise within firms on account of engineering or managerial considerations; external economies may be obtained by the geographical concentration of activities, so that services and other facilities may be shared in common. Both will lead to the geographical concentration of activities.

With the formation of the Common Market and the European Free Trade Area after the Second World War, and with successive rounds of negotiation under GATT, there was a widespread feeling in the 1950s and 1960s that considerable scope was available to achieve scale economies which had thus far not been realized. To the extent that this might occur, there was clear potential for the geographical concentration of activities in a manner which seemed to explain the actual pattern of prosperous and depressed regions. This fact intersected with the Keynesian emphasis on disequilibrium processes, giving scale economies considerable status as an important mechanism, allied with multipliers and the Verdoorn effect, creating cumulative growth processes.

Figure 4.2 illustrates the way in which production costs are usually portrayed in relation to scale of activity. Note, however, that there is an infinite array of possible shapes for the cost curve. Note also that it is generally believed that after a certain scale is achieved diseconomies will set in, giving a U-shaped curve, such as is illustrated in an inverted manner in Figure 3.4. Diseconomies of scale put an upper limit on the growth of a region. However, in the Keynesian model of regional development, it has generally been assumed that

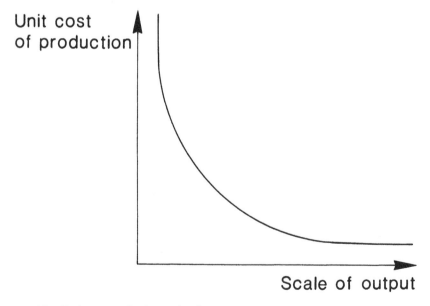

**Figure 4.2**   Scale economies in production

the curve is of the type shown in Figure 4.2; if diseconomies do exist, they are assumed to lie beyond the scale of development actually achieved. On this assumption, scale economies will lead to the continuation of cumulative growth (decline) once it has become well established.

## Policy conclusions

Keynesian analysis leads to policy conclusions relevant for regional affairs at two different levels. First, macro-economic stabilization measures will have a distinct impact on regions, in a manner which should reduce unemployment differentials compared with the situation without intervention. Second, to the extent that these effects are inadequate, high levels of regional unemployment are interpreted as due to a deficiency of demand, to counter which specifically regional policies are called for. We will briefly discuss the former before turning our attention to the latter.

Progressive income taxes and income support measures play an important part in macro-economic stabilization. If incomes are rising rapidly, fiscal drag ensures that an increasing proportion is collected in taxes; the government is not obliged to spend all of its increasing revenue, and if it does not, the national level of savings can be raised. The effect will be to dampen the boom. Conversely, at times of high unemployment, income support measures (especially unemployment benefit) will maintain incomes and hence expenditure at higher levels than would otherwise be the case. With demand maintained, the economic downturn will be ameliorated. These automatic stabilizers may be supplemented by counter-cyclical public expenditure on infrastructure – roads, airports, sewers, etc.

These instruments of macro-economic management have clear regional implications, tending to dampen boom conditions in prosperous regions and to ameliorate circumstances in those where unemployment is substantial. However, nobody seriously expects that these measures, on their own, would be sufficient to equalize regional unemployment rates, even in conditions of full employment at the national level. Residual pockets of higher unemployment may be interpreted as the consequence of inadequate demand for the products of particular industries located in the depressed areas. The existence of localized unemployment implies that total output in the economy is less than it might be, and also that adverse cumulative processes may be set in train in the afflicted areas. This being the case, intervention is necessary.

One form of intervention is ruled out, since it would violate basic premises of Keynesian thinking. It is assumed that in face of falling demand for labour, wages will be 'sticky', i.e., will not adjust downwards with any speed. This assumption derives from the belief that collective wage agreements create rigidity in this aspect of the economy, and more generally that price adjustments do not form part of the response processes in the economy. More important, perhaps, is the view that were wages to fall, regional income would also fall, which would have adverse multiplier effects on the regional economy. Consequently, intervention to secure a reduction in wages is ruled out as both impractical and undesirable.

The 'pure' Keynesian response to the existence of regional unemployment at a time when the national picture approximates full employment may be characterized as follows. Structural unemployment in the less fortunate regions arises from the decline of localized staple industries – coal mining and cotton textiles, steel or shipbuilding, as the case may be. Such a decline can be regarded as due to a deficiency of demand for the products in question, over which the government has no control. Therefore, if full employment is to be regained in the depressed regions, new demand must be injected (MacKay, 1982). Consequently, the problem becomes one of shifting demand from the prosperous regions to the less prosperous ones: 'Relocating industry is the typically Keynesian response to the problem of inadequate demand for labour in specific regions' (Wadley, 1986, p. 56). In the British case, where policies designed to effect the geographical reallocation of labour have been of minor importance: 'The complementary policy is to improve the matching between the demand and supply for labour by diverting the demand for labour to areas of high unemployment' (Armstrong and Taylor, 1985, p. 194).

To divert demand for labour from one region to another requires explicit policy instruments at the micro-economic level. In principle, such instruments may be of two kinds – those which operate directly on firms whose planned investment is deemed to be 'mobile', and those which serve to improve the operating conditions, and hence profitability and growth, of all firms (or a specified subset) in the depressed regions.

Administrative controls provide one mechanism whereby 'mobile' investment can be steered to particular regions. For many years in Britain, the building or extension of a factory could only be undertaken in the more prosperous regions if an Industrial Development Certificate had been obtained. It was common for such certificates to be withheld, especially at times when the pressure of demand in the prosperous regions was high. A rather similar strategy has been employed in Italy, the state industries having been required to locate 40 per cent of their investment in the Mezzogiorno. As an alternative to administrative controls, individual firms may receive financial assistance (usually with investment) on a selective, negotiated basis. Such assistance, available only in specified areas, provides an incentive to expand in or relocate to the assisted area.

Administrative controls and selective assistance introduce uncertainty and delay, to combat which an alternative or complementary approach is to provide assistance 'as of right' to qualifying firms, whether or not they are immigrants to the area or local businesses. The overall aim is to expand economic activity in the assisted areas in preference to other regions. Standard grants to assist with capital expenditure have been the most common measure, supplemented by: labour subsidies, operating subsidies (especially on fuel and transport costs), and the provision of infrastructure. The last of these includes the building of advance factories as well as transport facilities, public utilities and the like.

Most European countries in the post-war era have implemented policies based on some combination of the above measures (Yuill and Allen, 1980-).

Over time, the emphasis has shifted somewhat, away from administrative controls and toward financial inducements directed at firms, especially assistance with capital costs which, in the parlance of the European Community, are 'transparent', i.e., clearly identifiable and measurable.

Two related features of the Keynesian policy prescription deserve notice. Some commentators regard subsidies to firms and the construction of advance factories as constituting a supply-side approach to regional policy. At one level, this clearly is true, in that policy is operating through the cost of factors of production. At another level, though, such a description obscures the fundamental aim, which is to shift the focus of demand from one region to another. The *goal* is the management of demand, although the *means* employed are supply-side in character. The fact that supply-side means are used to further demand-management policies implies two things. First, it implies that the Keynesian assumption that supply will look after itself if demand can be got right is not in fact tenable. Second, it demonstrates that the antithesis between supply-side and demand-management approaches is, in part at least, artificial. Another and related point confirms the artificiality of the distinction. If a regional policy package based on capital or other subsidies, or infrastructure investment, succeeds in raising economic activity in a region above the level otherwise expected, how much of that increase represents diversion from other regions and how much represents extra activity which would not have existed but for the assistance provided? Only the former component represents the true spatial reallocation of demand resulting from the policy package; the latter amounts to a supply-side stimulus to the national economy.

The boundary between the manipulation of demand and the supply-side effects is clearly a difficult one to draw, the more so as supply-side means are employed to achieve demand-management ends. Nevertheless, to retain clarity of analysis, these distinctions ought to be maintained. To argue, as Armstrong and Taylor (1985, p. 194) do, that policies designed to transfer industries geographically should be defined to 'include policy instruments which induce the growth of indigenous economic activities *within* regions' serves to confuse rather than to clarify matters.

The disequilibrium, cumulative causation model of regional growth postulates that favoured regions will inexorably and inevitably forge ahead, while other regions decline in relative and perhaps even absolute terms. This positive theory suggests that if a government employs measures to concentrate new investment in a limited number of centres it will be possible to create the multiplier, scale and Verdoorn effects which will ensure continued self-sustaining growth (Rodwin, 1963). In the regional planning domain: 'The doctrine of growth centres became the normative counterpart to the positive theory of polarized development ... Regional planning doctrine in the 1950s and 1960s revolved essentially around the idea of growth centres' (Friedmann and Weaver, 1979, pp. 125 and 172). The extensive literature on the growth centre concept, and the somewhat arcane distinctions between growth centres and growth poles, has been well summarized by Hansen (1972) and Moseley (1974).

Figure 4.1 illustrates the policy principle. At any given time, certain industries may be regarded as 'propulsive', in the sense used by scholars such as
Rostow and Perroux. In the 1960s, the vehicle industry, steel and chemicals
were widely regarded as exemplary cases – output was rising strongly, technological development was occurring rapidly, and linkages (backward and forward)
were important features. To provide modern facilities and to cater for rising
demand, new capacity in these industries was being planned and built. Consequently, it was argued, some of the new investment in these and comparable
industries could be steered to suitable sites to provide the initial stimulus
required for establishing a growth pole, which would stimulate the economy of
the laggard region. In other words, the relocation of some existing demand,
combined with judicious initial expenditure on infrastructure, etc., could
trigger long-term cumulative growth which would not require continuing
public subsidy.

Practical expression of this idea was to be found in the British government's decision in the 1950s to divide a proposed new steel mill into two
(sub-optimal) units, one located at Newport in South Wales and the other at
Ravenscraig, near Glasgow. At the same time, efforts were made to disperse
the vehicle industry from the Midlands and South – most notable being the
move of Rootes to Glasgow. Soon after, with the publication of a White
Paper in 1963, eight growth centres (actually, areas) were designated for
Scotland. Similar policies have been pursued in Italy since 1959 and a year or
two later in France, with considerably more energy and commitment than has
been evident in Britain, though with a degree of success which has been below
expectations (Dunford, 1988). Elsewhere, the initial enthusiasm for growth
centre strategies has also waned in the light of experience (Friedmann and
Weaver, 1979).

## Conclusion

Neo-classical and Keynesian prescriptions for regional policy, derived from
very different assumptions and theoretical structures, are dramatically different
one from the other. For much of the post-war period up to the mid-1970s, the
neo-classical view was regarded as less plausible than the Keynesian formulation, for explaining what actually happens and in providing appropriate policy
guidance. Nevertheless, neo-classical ideas did not die and wither, for two
distinct reasons. First, in terms of theory, one can construe the neo-classical
model as a normative statement of how the world should operate, whereas the
Keynesian approach provides a positive framework of the way regional growth
actually occurs. In the second place, some of the available empirical evidence
did seem to support the neo-classical viewpoint, while as the 1960s melted into
the 1970s, experience with Keynesian policies suggested that they were more
fallible than had been expected.

Although empirical experience cannot provide an incontrovertible test by
which to assess the utility of competing theories, some appeal to reality is now
in order. Therefore, in Chapter 5 we will examine the light which experience can

throw on the value of the neo-classical and the Keynesian modes of thought and policy prescriptions for regional development. If we may anticipate the outcome of this discussion, we will note that actual experience suggests that both the neo-classical and the Keynesian formulations are, in important respects, wanting.

# CHAPTER 5

# *Neo-classical and Keynesian theories of regional growth in the light of experience*

> Theoretical disagreements of this fundamental kind are not really suscept-ible to empirical test. For even where the protagonists can agree on which are the crucial variables to be tested and how to match the theoretical concepts with empirical constructs, the economic system does not stand still.
>
> (Deane, 1978, p. 222)

During the reconstruction period after the Second World War, regional prob-lems were of minor importance compared with national needs to switch from wartime to peacetime production, to provide for new capital investment and an adequate supply of consumer goods. The resulting level of activity was, in most countries, sufficient to ensure that all regional economies were working flat out. The main exception among the advanced western nations was Italy, where the low income and high level of unemployment in the south proved to be fertile ground for the Communist appeal to revolution; as a directly political response, the Cassa per il Mezzogiorno was established in 1950, charged to make a large and quick impact on the poverty, initially by agricultural improvement and infrastructure investment. Thereafter, other countries adopted an increasingly interventionist stance over the regional disparities which began to re-emerge, first in Britain in the late 1950s and then elsewhere in Europe. By 1970, all the then member nations of the European Community were operating a package of regional policy measures cast in the Keynesian mould. In terms of practical policy, the neo-classical school of thought had been pushed into the back-ground. Nevertheless, some work continued to be published which suggested that interregional adjustments did occur in the manner postulated by neo-classical theory.

In this chapter, we will start by discussing some of the evidence favouring the neo-classical view of regional processes. We will then turn to a consideration of experience in operating Keynesian regional policies. In both cases, the evidence suggests that the real world behaves in a manner which, at best, is only partially explained by either of the two bodies of thought. But, precisely because 'the economic system does not stand still', this evidence is suggestive, and provides no certain 'proof'. However, we will be in a better position to understand why, in the 1970s and 1980s, there has been a widespread sense of disenchantment with both schools of thought.

## The neo-classical case

Two influential studies were published in the mid-1960s, both of which seemed to provide powerful support for the idea that in advanced nations the equilibrating mechanisms postulated by neo-classical theory actually work (Borts and Stein, 1964; Williamson, 1965). Williamson explored the evolution of regional income differences as economic development proceeds, using cross-section data for several countries, and time-series information for the limited number of nations for which data existed. He used the coefficient of variation as his measure of income disparity, with and without weights for the size of the regional populations. His results display a consistent pattern, in which regional income disparities become greater as development proceeds from a low base, reach a peak and then, in the advanced (high income) nations, diminish. Of the twenty-four countries included in his cross-section study of data for the late 1950s or early 1960s, eleven could be described as advanced. Of these eleven, nine displayed disparities far below those in any of the poorer nations. His time-series data relate to the more advanced countries, for the most part. Generally, regional income disparities became much less marked over the period from the 1930s to the 1950s.

Williamson reported his empirical findings with minimal discussion of the mechanisms which might give rise to the observed patterns. His general hypothesis is formulated in the terms used by Myrdal and Hirschman, that in the initial stages of development there is a geographical polarization which then gives way to the spread of economic growth across the national domain. For the advanced nations, his data strongly suggested that the latter part of the Myrdal-Hirschman thesis is valid. The neo-classicists would claim that, given the initial disequilibrium of polarized development, equilibrating mechanisms would come into play.

Borts and Stein address these issues directly, using the United States as their laboratory, over a period from *c* 1880 to 1949–51. Their starting point was the clear evidence of income convergence between the states, as measured by the coefficient of variation (unweighted). First, they test whether a one-sector neo-classical model provides an adequate explanation of events. Their conclusion is unequivocal; it does not. Thereafter, they essay a number of analyses using disaggregated data, leading to the conclusion that income convergence can be explained by a combination of intra-state transfers between sectors, especially the reduction in the importance of agricultural employment, and transfers of labour and capital between the states. The growth of the labour force appears to have been a major determinant of the development patterns. Overall, they conclude that: 'the US interregional and interindustrial growth pattern seems to be tending towards a competitive equilibrium and hence towards intertemporal efficiency' (Borts and Stein, 1964, p. 214).

However valid these studies may have been for the countries concerned and the time periods analysed, within a decade of their publication events appeared to show them as inapplicable in current circumstances. In many of the advanced countries, unemployment levels began to edge upwards in the late 1960s, and then rose sharply in the 1970s. At the same time, inflation took off and regional differentials (of unemployment and income) began to widen. Initially, these changes were perceived as short-term, temporary aberrations from the controlled prosperity of Keynesian economic management. By the mid-1970s, the realization was dawning that something 'permanent' had happened. The reality of stagflation was as damaging for the neo-classical view as it was for the Keynesian model. Even so, neo-classical processes can still be detected, as, for example, in the relationship between wage increases and changes in the level of unemployment in Great Britain, measured at the *county* level in England and Wales and regions within Scotland – a finer spatial disaggregation than the standard regions and a closer approximation to local labour markets. Even though the absolute level of wages correlated only rather weakly with the level of unemployment for both the periods 1978 to 1980 and 1981 to 1983, changes in wage levels for males showed a reasonably good relationship with changes in earnings. The greater the increase in unemployment, the lower the level of wage increases relative to the national norm (Bentham, 1985).

The implications of high and rising unemployment for regional policy were explored by Chisholm in 1976, since when there has been a very considerable literature, some of which we will review later in this chapter and in Chapter 6. The United Kingdom, for example, has experienced a widening of regional income differentials during the last decade and a half, as well as greater absolute differentials in unemployment rates (Martin, 1988a). The Commission of the European Communities (1984, 1987) has compiled evidence that whereas up to 1979 there was a tendency in several European countries for regional income disparities to become less marked, since then differentials have widened. Thus, as seen from the vantage point of the 1980s, it is by no means clear that regional convergence is the normal process in advanced countries, a fact which seems to negate the neo-classical view. However, the evidence offered by Williamson and by Borts and Stein is essentially long term in nature – decades and demi-centuries. While it is clear that the neo-classical model does not give a *continuous* process of convergence, we cannot conclude that equilibrating tendencies are *wholly* absent. This fact becomes particularly clear if we consider the evidence pertaining to migration, a matter to which we turn in the next section.

## Modelling movement

There is a very large class of phenomena which may be labelled as movement, ranging from the daily patterns of people's work, shopping and leisure trips, to the regular despatch of goods and permanent migration. The great majority of daily movements from the home base occur within the boundaries of regions and are therefore not of direct interest in the present context. Nevertheless, these movements are equilibrating, in that they arise from the spatial separation of activities and make a stable pattern of such separation possible. In the present context, however, it is the interregional shifts which are more important, since these shifts should, if neo-classical theory is correct, tend to bring economic systems which are in disequilibrium into equilibrium.

There has been a large literature concerned with the theory of interaction models (see for example, Alonso, 1978; Rogerson, 1984; Wilson, 1971, 1980), in part reflecting the fact that there is a large family of related models. Stated in general terms, the amount of movement between two regions ($M_{ij}$) is some function of the 'sending propensity' ($G_i$) of location i, of the 'attraction' exerted by location j ($K_j$) and of the separation effect which arises from the distance between i and j ($d_{ij}$). In Mueser's (1989) terminology:

$$M_{ij} = G_i K_j d_{ij}$$

To convert this basic expression into operational terms, it is necessary to specify the characteristics of $G_i$ and $K_j$. For example, in a gravity model the propensity to send and to attract is often taken to be a function of mass, or population size. However, depending on the purpose of the study, it may be relevant to incorporate levels of unemployment, wage levels or environmental qualities as elements in the measurement of $G_i$ and $K_j$. Similarly, for $d_{ij}$ one might take distance in kilometres, the cost of the journey or the time it will take. Once these decisions have been taken, consideration must be given to the constraints which are relevant, for example, that all movements which originate must be accounted for at the destinations, or that particular links in the movement network may have capacity limits. If the exercise is to determine an optimal pattern of movement, one will wish to specify the objective function – the property which is to be maximized or minimized. If, on the other hand, the purpose is to model reality (that is, positive analysis), then the task becomes one of specifying and calibrating the scalars for the variables.

Interaction models in the above form clearly have the potential to show whether actual patterns of movement do accord with the neo-classical view of the world. The main application relevant to us has been in the field of migration; this literature has been well summarized by Greenwood (1975) and Willis (1974) for the United States and by Woods (1979) for the United Kingdom. Greenwood (p. 397) begins his review by referring to J. R. Hicks' dictum, that 'differences in net economic advantages, chiefly differences in wages, are the main causes of migration'. The post-war evidence from the United States does not support that assertion if wage differences are defined narrowly in money wage terms, even if wages are adjusted for differences in cost of living. However, if net economic advantage is taken to include the availability of amenities, access to friends and relatives and environmental (climatic) variables, then it appears that quite high levels of explanation can be achieved.

American evidence strongly suggests that a distinction must be made between the way we approach the modelling of origins and of destinations. The age, domestic status and skill of individuals has a material impact on whether a potential migrant does actually move, thereby masking the effect of income and job opportunities in the origin area. On the other hand, once individuals have elected to move, their choice of destination is strongly influenced by these 'economic' variables, if these are interpreted widely. Migration flows do, therefore, represent an adjustment process which has some elements of the equilibrating mechanism postulated in neo-classical theory. However, the distance exponent, which forms a major focus of research (e.g., Mueser, 1989), remains something of a problem. The reality of distance-decay is not in doubt, though estimates of its magnitude vary. The problem, perceived in strictly economic terms, is that the rate at which migration decreases with distance generally exceeds the objective cost of translocation. Therefore, some part of the $d_{ij}$ function must be interpreted as arising from some combination of psychic costs and information costs. Given that personal characteristics affect the propensity to migrate (hence, the pattern of origins) and given that the distance exponent exceeds the objective cost function, migration streams will be only an imperfect mechanism leading to equilibrium between regions.

Personal characteristics influence migration streams in more complex ways (Stillwell, 1978). The age, sex and ethnic characteristics all have a bearing on both the propensity to migrate and on the responsiveness of individuals to the attractions of potential destinations. One of the more thorough attempts to model this complex system is Congdon's 1989 study of the boroughs which comprise Greater London. He used two sets of data, one for the period 1971 to 1981 and the other for the year 1980–1, to analyse the causes of interborough migration (not the effects). The analysis includes data on unemployment rates, employment growth, incomes, house prices and housing tenure, and occupations. He found that the:

> migration data over the entire period 1971 to 1981 suggests that migration between boroughs within the metropolis plays an equilibrating role in housing and labour markets, but is not necessarily simultaneously equilibrating in both. For example, high house prices may act to deter immigration to more prosperous boroughs with higher economic growth.
>
> (Congdon, 1989, p. 101)

Intra-national studies, therefore, provide some evidence to support the neo-classical view of migration as an equilibrating mechanism and thereby support Thomas' (1954) classic study of international migration across the Atlantic, in which he found that fluctuations in movement could be associated with demographic changes and shifts in investment patterns. On the other hand, the expected associations are not always very strong. Equally important, to the extent that migration does serve as an equilibrating role, it only does so over quite long periods of time. Therefore, an initial disequilibrium cannot be eliminated quickly; indeed, it is more likely that the disequilibrium will be reduced than that it will be caused to disappear. However, studies using aggregate data ignore the selective aspects of migration – by age, sex, education and skill. Even if aggregate flows fit the equilibrating model postulated in neo-classical theory, there may be long-term disequilibrating effects as the population structure in both the sending and receiving regions is altered.

Finally, the American evidence points to the role of migration as the cause of economic development, rather than as a response to the opportunities created by development. This is implicit in the role of the environmental factor, as noted by Greenwood (1975), and has been remarked explicitly by some authors (for example, Wheaton, 1979a). This is a point to which we shall return in the context of cumulative causation (p. 93).

Much less work has been done on interregional movements of capital. The outstanding exception is Engle's (1974) study of Massachusetts, over the period 1956 to 1971. Four sectors of manufacturing were identified. In each case a comparison could be made between the level of profitability in Massachusetts and in other states within the United States. Engle found that the rate of investment in Massachusetts varied as a function of the relative profitability of the Massachusetts industries, and his explanatory equations yielded high levels of $R^2$. If the rate of profit in Massachusetts rose by 1 per cent, or if it fell by the same amount elsewhere, investment within Massachusetts would rise by between 0.6 per cent and 1.6 per cent, depending on the industry:

| | |
|---|---|
| Textiles and apparel | 0.64 |
| Electronics-machinery-technology | 0.86 |
| Primary and fabricated metal products | 1.50 |
| Printing and processed foods | 1.57 |

These elasticities were calculated to allow for a three-year adjustment period following a change in relative profitability. Numerous factors affect profitability – input prices, wages, labour productivity, etc. – but these were not investigated. The key point is this: given that profitability differences exist, the flow of investments does adjust in the manner predicted by neo-classical theory, but that it does so differentially by industrial sector, and with a distinct time lag.

## Conclusion

At best, the neo-classical model provides an incomplete account of what has actually been happening, an account which fails to capture important parts of the overall process of regional change and adaptation. The useful insights which it does offer are relevant to rather long periods of time – decades and demi-centuries – rather than for the shorter periods customarily associated with interventionist policy. Clearly, the impediments to adjustment are considerable, a fact which puts in question one of the basic premises of the neo-classical model. Because the neo-classical view of the world is not wholly correct, the temptation may be to regard it as wholly useless. More realistically, we need to keep an open mind regarding the (limited?) utility of the neo-classical approach until we have explored further.

## The Keynesian case

As we have already noted, most European countries adopted regional policy packages after the Second World War. Not unnaturally, governments and scholars were curious to know how effective these policy measures would prove to be. Therefore, we will begin our enquiry by reviewing the evidence concerning the effects which have been attributed to the measures taken by governments. Thereafter, we will consider what has been learned about the magnitude of regional multipliers and the strength of linkages between firms; the impact of policies on indigenous firms as distinct from immigrant firms; the operation of cumulative causation, including growth poles; and the impact of demand management policies on one particular industry, motor manufacturing, in Britain.

### *Evaluation of regional policy packages*

Broadly, there are two routes which may be followed in evaluating regional policy packages. First is the use of macro-models, which attempt to measure shifts of aggregate employment or investment between regions. The alternative approach is to engage in micro-studies of individual firms, commonly by enquiring into the reasons for investment, location and manning decisions. Unfortunately, these two approaches yield conflicting conclusions: in general, it appears that the macro-approaches to estimating the effects of policy show a greater responsiveness of firms than is manifest in micro-studies (Nicol, 1982). Given that most of the well-known evaluation exercises are of the former kind, there may be a bias in the general view concerning the effectiveness of policies.

Ashcroft (1982) assembled the available evaluation studies for six European countries and found that: 'The results are largely favourable to regional policy' (p. 302). For example, in the United Kingdom it appears that one or more of the assisted areas had between 16 and 25 per cent of its industrial capital attributable to the policy package; in Germany, the equivalent range was 6 to 15 per cent. On the other hand, the number of jobs associated with policy measures proved to be rather small. Job creation can be treated as the number of jobs per annum for every thousand of the population. The estimates cited by Ashcroft are: 1.19 jobs in the United Kingdom, 1.23 in Ireland and 1.26 in Germany. In this respect, the United Kingdom is particularly instructive. Between 1960 and 1976, about 0.48 million surviving jobs had been 'created' in the assisted areas (Armstrong and Taylor, 1985, p. 284). However, for the somewhat longer period from 1961 to 1981, the net gain was estimated to be only 0.37 million jobs (Moore, Rhodes and Tyler, 1983, p. 11). This discrepancy may be explained by the 'plateau effect', namely, that with the passage of time some firms which had moved to assisted areas closed down. To an increasing extent, therefore, the gross effect was being offset by the loss of jobs which had previously been 'created' by policy. Therefore, in order to maintain a given annual net addition to employment in the assisted areas, the scale of assistance would have to rise steadily.

Although the macro-studies show a positive impact attributable to regional policy, it had become clear by the early 1980s that this impact had been insufficient to 'solve' the regional problems, and that no solution was in sight, even if economic conditions had not taken a sharp turn for the worse. Consequently, experience from the 1960s and 1970s held out little hope that, once the 1979–83 recession was over, a continuation of the regional policy measures would achieve the desired aims.

These doubts are compounded by the intractable problem of disaggregating the policy packages, to identify the effect of individual instruments. Clearly, it is desirable to know how cost-effective instruments are, so that resources may be concentrated upon those which achieve the most effect for least cost. Unfortunately, little confidence can be placed on the disaggregated effects (Ashcroft, 1982; Nicol, 1982). Perhaps the extreme case is provided by the operation of the Industrial Development Certificate (IDC) policy in the United Kingdom. The ratio of IDC refusals to the number of applications was a widely used measure of the 'strength' of regional policy, a high refusal rate being an indication that policy was being applied strictly. The general expectation was that firms refused a Certificate in the more prosperous areas would locate in one of the assisted areas or in a new town. As Table 5.1 shows, that expectation was not matched by the actual decisions of the firms which were refused an IDC.

Overall, the Exchequer cost per job for the jobs 'created' in the British assisted areas over the period from 1961 to 1981, at 1981 prices, amounted to £32,000, plus a total additional expenditure of over £3 billion to provide government-built factories and other industrial infrastructure (Moore, Rhodes and Tyler, 1983, p. 15). Not surprisingly, the government felt that these costs were excessive and introduced measures designed to curb the upward trajectory of regional policy expenditure (Chisholm, 1985b; Martin, 1985). One reason for wishing to shift the emphasis towards more 'selective' assistance, and away from the existing investment grants given as of right, was the accumulating evidence that much of the expenditure was going to indigenous firms, rather

**Table 5.1**  United Kingdom: the location decisions of firms which were refused an Industrial Development Certificate, 1958–71

| Location decision | Percentage of cases analysed |
|---|---|
| 1   Project was located in an area acceptable to the government (e.g., an assisted area or new town) | 18 |
| 2   Project located in a non-assisted area but in a modified form (e.g., located in existing buildings) | 50 |
| 3   Project was located abroad | 1 |
| 4   Project was abandoned | 13 |
| 5   Other (e.g. firms subsequently closed or reorganized) | 18 |
| | 100 |

*Source*: Armstrong and Taylor, 1985, p. 205

than immigrant ones. Of the 0.37 million net jobs 'created' as at 1981, only 42 per cent were attributed to net movement; the other 58 per cent were jobs in local firms. It was far from clear that policy was achieving the relocation of demand which was its ostensible purpose. Equally important, it was evident that quite a lot of public money was being used to help finance projects which would have gone ahead anyway. Rather than influencing the geography of employment, such expenditure served only to increase the profits of the recipient firms. Possibly the most notable example is the Sullom Voe oil handling facility in Shetland, for which large grants were paid but which would have gone ahead in any case. In principle, of course, payments made under these circumstances might enable firms to reduce their prices, i.e., pass on the benefit, and thereby expand their business operation. No persuasive evidence has been brought forward to suggest that this has in fact happened.

Even without the severe recession from 1979, it had become clear that conventional regional policy packages were achieving rather little in relation to the expenditure incurred. Numerous authors began to call for better targeting of regional aid, and for the involvement of local authorities and other agencies, in an effort to make policy more effective (e.g., Martin and Hodge, 1983; Regional Studies Association, 1983). But why had the achievements of regional policy packages seemed so disappointing? Part of the reason lies in the fact that regional multipliers, and the strength of local industrial linkages, both turned out to be weaker than had been believed.

## *Multipliers*

If intervention to control the level of demand in a national or a regional context is to be successful, reasonably accurate estimates of the relevant income (or employment) multipliers are essential. At both the national level and regionally, there has been, and continues to be, uncertainty regarding the magnitude of multipliers. However, notwithstanding the uncertainty, accumulating evidence suggests that in practice the values are lower than had been supposed. Aldcroft (1984, pp. 39–40 and 51) makes this point with respect to the United Kingdom. At the time of Lloyd George in the 1930s, it was common to suppose that the income multiplier for the nation lay in the range two to three, and Keynes himself worked with a value around two. However, more recent work on the 1930s suggests that the national multiplier, calculated over twelve years, was in fact only 1.44. If that estimate is correct, then to reduce unemployment by 3.0 million, using Keynesian techniques, would have necessitated expenditure equivalent to 19.3 per cent of the 1932 GNP – a sum so large as to be politically and economically improbable to achieve. To have obtained a similar reduction in 1981–2 would have required expenditure of £47 to £48 billion, approaching 15 per cent of GNP.

Estimates for the size of regional multipliers have also been scaled down. Writing in 1973, Gordon noted:

> Over the last 5 years or so, there has been a considerable growth of regional multiplier studies in this country [Britain] ... the resulting esti-

mates have virtually all been low by comparison with the expectations aroused by some earlier discussions of regional policy issues.

(Gordon, 1973, p. 257)

The accuracy of this statement can be readily verified. Working with exiguous data, Archibald (1967) estimated that multipliers for the British standard regions lay in the range of 1.2 to 1.7. Two years later, with much improved data to hand, Steele (1969) obtained the following results for multipliers with feedback:

> Lowest multiplier: Yorkshire and Humberside   1.26
> Highest multiplier: Scotland                              1.77 or 1.92

Subsequent reworking of the data, however, yielded somewhat lower figures (Steele, 1972):

> Lowest multiplier: West Midlands            1.14
> Highest multiplier: Southeast                  1.50

Steele's revised figure for Scotland (at 1.4) is identical to that obtained by Allen (1969). However, another set of estimates by Brown (1972) put regional multipliers in the range from 1.15 to 1.24, markedly lower still. Two other studies confirm the relatively small size of regional multipliers in Britain. Archer (1976) quotes a tourism study for Tayside which yielded a multiplier for the whole of Scotland of 1.46, while the range of sectoral multipliers has been estimated for North Staffordshire by Pullen and Proops (1983). For 27 sectors, they obtained 'value' multipliers as follows:

> Lowest multiplier: national government             1.010
> Median multiplier: medical and dental services  1.099
> Highest multiplier: agriculture                          1.398

The second highest multiplier was 1.292, for the bricks, pottery and refractory goods industry – the long-established staple industry of Stoke-on-Trent. North Staffordshire is, of course, a smaller region than the standard regions we have otherwise considered. Nevertheless, their conclusion seems to accord with the general run of findings: 'The regional multipliers for the various sectors are all quite small, so the indirect effects of exogenous shocks to the export base are predicted to be small relative to the direct effects' (Pullen and Proops, 1983, p. 199).

The empirical studies have, therefore, confirmed the scepticism of T. Wilson (1968), who published one of the early attempts to assess the utility of regional multipliers for regional policy. Summing up his discussion, he noted:

The first of these questions related to the possibility of reducing regional disparities by stimulating expenditure in the development areas without having any significant spillover into regions already fully employed. In this case a definite enough answer can be given, and it is in the negative.

(Wilson, 1968, p. 390)

A similar scepticism had been voiced by W. R. Thompson (1965, pp. 205–6) for the United States:

> The local multiplier is much lower than the national multiplier, probably ranging from a value of perhaps one-half as large in a million plus population urban area down to perhaps one-quarter as large in an urban area of 50 thousand population. Unlike the national case where the multiplier must be at least one and is probably rarely less than two, the local multiplier could easily be less than one.

We would expect the national multiplier to be larger for the United States than for the United Kingdom, because the United States is a much larger and more nearly self-sufficient economy. However, if a regional population of about 1.0 million would experience a multiplier of about unity, it seems that regional multiplier effects in both countries are actually quite low.

Two main reasons for the low value of regional multipliers are worth noting. First, central government taxation (personal taxes and taxes on expenditure, such as VAT) ensures that a substantial proportion of any extra local income is syphoned out of the region – possibly as much as 40 per cent in Britain. In the second place, out of the residual income available for expenditure, a large proportion is spent on commodities and services which are purchased from outside the region. The fact that the propensity to import is both high and probably rising should occasion no surprise, given the evidence concerning trends in international trade discussed in Chapter 2. One part of this propensity to import is accounted for by the nature of industrial linkages.

*Industrial linkages*

In the early post-war period, it was widely accepted that the long-established Victorian industries had become highly localized for a combination of reasons, which included the external economies available to firms by locating in close proximity to each other. These economies included the availability of local specialist suppliers of machinery, components, etc. This feature seemed to be widespread among industries such as shipbuilding and textiles (Florence, 1948, 1962; Smith, 1949), although Allen (1959) notes that shipbuilding firms purchased some inputs from abroad or from Birmingham. Nevertheless, the general image of the older industries, of close local linkages, was widely held to be the economical way in which to organize industry, a view that received confirmation in studies of Birmingham's complex manufacturing economy (Allen, 1929; Wise, 1949) and of London's clothing and furniture trades (Hall, 1962). The pattern seems to have survived in some 'traditional' kinds of industry such as iron foundry, though more notably in East Lancashire than in the West Midlands (Taylor, 1973).

As governments became increasingly interested in the scope for relocating industries, the question naturally arose regarding both the transition costs incurred and the long-term, ongoing cost differentials for plants in different locations. In particular, if branch plants would incur higher costs in locations remote from the parent company, relocation policies could be damaging to long-term prosperity. Preliminary studies of sample firms suggested that there was in fact considerable scope for relocation, at least in selected industries (Hague and Newman, 1952; Luttrell, 1952, 1962), if firms adopted sensible strategies regarding the managerial system used. But the case studies did suggest that finding local suppliers of inputs to serve branch plants could be a problem. Nevertheless, it was generally believed that with the passage of time local sourcing could be established and this was certainly the view of successive governments in Britain in the 1950s and 1960s. This was especially clear with the move of the vehicle industry to Merseyside and the establishment of both steel and vehicle manufacture in the Glasgow area. In both cases, it was expected that local linkages, as postulated in growth centre theory, would quickly develop, so that the modern industrial sectors would replicate the linkage patterns established in Victorian times.

Keeble's 1969 study of northwest London was probably the first explicitly to point out that modern industries had linkage patterns which differ sharply from those established by the older (and declining) lines of manufacture. Northwest London was characterized by expanding engineering firms making a wide range of mechanical and electrical goods. Their forward linkages (that is, sales of products to other manufacturers) showed: 'that unlike the situation in inner London, local linkage has not played a dominant role in north west London's industrial organization and growth' (Keeble, 1969, p. 169). Of the firms sampled in his study, only 27 per cent supplied more than one-quarter of their output to other firms in the locality; half the firms had no local sales at all.

Other work has shown that the industries of north west London are fairly typical of industries in Britain and elsewhere in the weakness of their ties to other local firms. For a wide range of manufactures in Philadelphia (United States), the average proportion of inputs locally purchased (by value) was found to be only 19.4 per cent; no two-digit SIC industry even reached one-half (Karaska, 1969, p. 363). Just five years later, three separate studies – of Montreal, Toronto and Scotland – all showed that local linkages are weak, notwithstanding the relative isolation of both Montreal and Scotland in particular (Hamilton (ed.), 1974). A sample of firms located in Scotland showed that only 20 per cent of purchases were obtained from Scottish sources, although 76 per cent of sales were directed to that region (Lever, 1974, pp. 319–20).

For migrant firms, it may be expected that the pre-move pattern of linkages will initially be maintained but that over time new linkage patterns will evolve in which local purchases and sales will become more prominent. However, that expectation is not in fact supported by actual experience. From their analysis of two surveys, covering Greater London, East Anglia and the Northern region, Moseley and Townroe concluded that: 'The results clearly indicate that for

most migrant concerns the establishment of new local links is of little importance' (1973, p. 143).

More recent work confirms that local industrial linkages are generally rather
weak. For the assembly of motor vehicles in Merseyside, only 5.5 per cent of
intra-firm shipments in 1976 originated in the Northwest region, almost two-
thirds being shipped from the West Midlands and the Southeast (Stoney, 1986).
Electronic 'original equipment' manufactured in Scotland used only 19 per cent
of components and subcontract inputs derived from within Scotland in 1979,
the rest being obtained outwith that country (Firn and Roberts, 1984, p. 306).
Other studies of high technology industries generally show quite low local
linkages in both Britian and the United States (e.g., Keeble, 1989; Macgregor *et
al.*, 1986; Oakey, 1981). In her survey of three major high technology plants in
the United States, located in Arizona, Florida and Texas, Glasmeier (1988)
found that local sourcing accounted for only 5 per cent of manufacturing
supplies in the first case and that in the second case less than 1 per cent of
supplies were obtained from within the state. No figure is given for the Texas
plant, but local sourcing was reported to be negligible. On the other hand, small
firms do seem to be more reliant on local sourcing than are larger firms. Even
so, a sample of high technology firms in the San Francisco area, in Scotland and
in southeast England shows that only about one-half of firms purchase more
than 50 per cent of their inputs within a radius of 48 kilometres, and only
one-third dispatch a like proportion of their sales within that radius (Oakey *et
al.*, 1988, pp. 86 and 94). The San Francisco Bay area (which includes Silicon
Valley) diverges from this average; three-quarters of its firms purchase more
than half their inputs locally, although the sales links are similar to those which
exist in the two British regions. Regional comparisons are somewhat hazardous, however, on account of the small sample size.

One of the variables that will have an important impact on the strength of
local input linkages is the nature of the products being purchased. In this
respect, the findings of Hagey and Malecki (1986) are particularly instructive.
They examined four high technology industries in Florida and found that
purchases within Florida were, overall, a rather small proportion of total input
purchases. However, whereas barely one-quarter of firms purchased more than
20 per cent of sophisticated inputs within Florida, just over half of the firms
bought more than 20 per cent of routine inputs from within the state. In other
words, for those inputs requiring careful dovetailing of products (and hence
personal contact and quite possibly Research and Development), proximity is
less important than the ability to meet specifications.

Scott and Kwok (1989) examined the subcontracting behaviour of 33 firms in
southern California making printed circuit boards. On average, expenditure on
subcontract work amounted to 3.8 per cent of the final sales value, with a range
from zero to 24.9 per cent. The authors found some association between this
proportion and the generalized accessibility of the firms to the population of
subcontractors. The more accessible the firm, the greater the amount of subcontracting, a result which is interpreted as signifying economies from
agglomeration. However, given the very small number of firms, the large
number of other variables and the absence of direct evidence on the sources of

subcontract work and of the profitability of the firms, the conclusion that major agglomeration economies exist for this industry must be accepted with extreme caution, not least because it runs counter to so much other evidence.

Much has been made of the re-emergence of agglomeration economies with the appearance of just-in-time and other elements of new flexible production, 'Toyota City' in Japan being a much-quoted example of the spatial proximity of subcontractors (for example, Estall, 1985). One might expect, therefore, that Japanese firms engaging in inward investment into other advanced countries would replicate the model of an industrial complex. In practice, such inward investment takes place in the context of the existing distribution of potential subcontractors. Morris (1989) examined experience with Japanese inward investment in Wales (electronics), Germany (video cassette recorders) and Ontario (vehicles), and found that linkages occur over quite large distances. His scepticism about the general emergence of new industrial complexes is confirmed by the British experience with electronic consumer goods, high fidelity audio equipment and domestic electrical appliances. Europe is sufficiently small, and transport systems are sufficiently good, for just-in-time and related practices to be adopted even though components are drawn from several countries (Milne, 1989).

Thus despite the *a priori* argument that flexible production systems imply the advantage of proximity, there is little clear evidence to suggest that in practice linkage patterns are now characterized by shorter links than hitherto. Thus:

'it seems safe to assume that agglomeration economies associated with local material linkage have little to do with the recent pattern of regional industrial change' (N. Marshall, 1987, p. 110).

## Mobile firms/indigenous firms

In Britain, as elsewhere, regional policy was predicated on the attraction of mobile firms away from areas of prosperity to areas in need of assistance. It came as a considerable shock when, in the late 1970s, two major studies showed that in fact regional changes not associated with movement were at least as important as changes linked with the mobility of firms. Over the period 1966 to 1974, Greater London suffered a net loss of 390,000 jobs in manufacturing industry, representing more than one-half of the national reduction, and equivalent to 27 per cent of the capital's industrial employment (Dennis, 1978). London's decline could not be attributed to adverse industrial structure. Since only 16 per cent of the job loss was associated with the movement of firms to the assisted areas and the new towns, and some 44 per cent was due to the complete closure of plants, with a further 30 per cent due to contraction, it was shockingly clear that London was not a national dynamo supplying large numbers of mobile firms which could be steered to other regions. There was something seriously wrong with the performance of the capital's economy at a time when regional policy was being pursued vigorously on the assumption that London was providing, and would continue to provide, an inexhaustible supply of mobile firms which could be attracted to regions in need of extra jobs.

**Table 5.2**  Great Britain, 1959–75: sources of employment change, as a percentage of 1959 employment

| Category of subregion | National component | Structural component | Differential shift | Industrial movement | Indigenous performance |
|---|---|---|---|---|---|
| London | −5.3 | +8.3 | −40.8 | −12.1 | −28.7 |
| Conurbations | −5.3 | −4.2 | −6.3 | +1.5 | −7.8 |
| Free-standing cities | −5.3 | +1.0 | +9.1 | +1.1 | +7.9 |
| Industrial towns | −5.3 | −0.4 | +21.9 | +4.2 | +17.7 |
| County towns | −5.3 | +1.3 | +32.8 | +7.3 | +25.5 |
| Rural areas | −5.3 | −1.6 | +84.1 | +29.4 | +54.7 |

*Source*: Fothergill and Gudgin, 1982, pp. 54 and 73

One year later, Fothergill and Gudgin published a study which was subsequently extended and rewritten in 1982. Using shift-share techniques, they found that over the period from 1959 to 1975, for Great Britain as a whole, the structural component was much less important than the differential shift in determining the growth of regions. That is to say, changes in employment arising from the industrial structure of the regions were less important than changes attributable to 'local' (regional) effects operating on the rate at which industries grew or declined. Furthermore, the relative importance of the differential effect had been increasing over time, and this effect was dominated by indigenous performance in all of the regions into which the authors divided Great Britain (see Table 5.2). In this table, the figures for industrial movement and indigenous performance sum algebraically to give the differential shift; in every case except one (the rural areas), indigenous performance was more than twice as important as industrial movement.

A more recent study, covering the periods 1971–8 and 1978–81, might be interpreted as indicating the reversal of this pattern (Owen *et al.*, 1986). For the great majority of the areas into which Britain was divided, the structural component dominated over the differential shift component. However, in both periods, but especially in the second, there was general contraction in industrial employment, so that the structural effect was dominantly negative. In contrast, although the differential effect was smaller, it was generally positive.

**Table 5.3**  United Kingdom, 1961–81: estimated industrial employment effects of regional policies in all development and special development areas, 000 jobs

| | Immigrant firms | | Indigenous firms | | Total | |
|---|---|---|---|---|---|---|
| | Net | Gross | Net | Gross | Net | Gross |
| Thousand jobs 'created' | 153 | 232 | 212 | 274 | 365 | 506 |
| Net jobs as % of gross | 66 | | 77 | | 72 | |

*Source*: Moore, Rhodes and Tyler, 1983, p. 11

The importance of indigenous change has been confirmed in the assessment of the British regional policy package. Over the period 1961 to 1981, indigenous development accounted for 58 per cent of the net additional employment in the assisted areas, attributable to regional policy (Table 5.3). By the early 1980s, it had become clear that conventional regional policy was probably misconceived in so far as it was intended to influence mobile firms' decisions; in practice, the decisions of indigenous firms are at least as important.

Most European countries have shared Britain's experience. In the editors' introduction to a collection of studies on new firm formation in Europe, Keeble and Wever note that:

> no region, least of all the problem areas, can rely any longer solely on an external solution for solving their employment problems.
>
> In consequence, regional policies are increasingly emphasising the need to harness the indigenous potential of the problem areas.
>
> (Keeble and Wever 1986, p. 2)

If the aim of policy is to shift existing demand from one region to another ('pure' Keynesian policy), then instruments which simultaneously provide assistance to both 'mobile' and 'indigenous' firms imply an element of expenditure which is in excess of that which is needed. Furthermore, that excess expenditure, on the 'indigenous' firms, is more accurately described as an enhancement of the supply potential of the economy than as the spatial manifestation of demand management. Unless, that is, the 'indigenous' firms would otherwise have migrated to the very areas from which policy was seeking to divert 'mobile' firms. In the absence of any evidence that this might be so, one must conclude that, in so far as a policy package assists 'indigenous' as well as 'mobile' firms, it ceases to be a pure Keynesian policy and becomes a hybrid between the demand-management and supply-side approaches.

## Cumulative causation

The cumulative causation thesis predicts the continuing growth of regions which, for some reason, have gained a head start, and also the possibility of modifying that pattern of development by deliberately fostering growth centres in regions which are lagging. It is instructive, therefore, to compare some of the experience and attempts at policy implementation against prior predictions. For this purpose, we have some particularly good documentation pertaining to the United Kingdom and the United States.

When the Royal Commission on the Distribution of the Industrial Population reported in 1940, it clearly had been much influenced by the idea of an axial belt of industrial concentration extending from London to Merseyside-Manchester and Leeds (Figure 5.1). The existence of such a belt had been suggested by the Royal Geographical Society in its evidence to the Commission (Royal Geographical Society, 1938). The idea had also been discussed in a paper given to the Society, summarizing that evidence (Taylor and others, 1938). Attention was drawn to the decline of export staples such as coal and

**Figure 5.1** Great Britain: axial belt of industry
*Source*: Taylor *et al.*, 1938

textiles, and to the rise of newer industries (e.g., electrical goods and vehicles). The latter were regarded as serving the home market, and of having limited ties to raw materials (indigenous or imported). Consequently, it was held that access to the home market was the primary location factor. Taylor and her colleagues clearly envisaged that, in the absence of countervailing action, industrial employment would become progressively concentrated in this axial belt.

The Royal Commission accepted the principle of cumulative growth and the reasons giving rise thereto, and broadly endorsed the idea of the suggested axial belt. However, the Commission particularly emphasized the attractive powers of London and the home counties. Furthermore, the Commission published data which cast some doubt on the reality of the axial belt as a fact of Britain's geography, their data indicating that the main area of cumulative growth lay in the London region of the Southeast. Notwithstanding those doubts, and a devastating critique by Baker and Gilbert (1944), the idea of an axial belt became quite widely embedded in thinking about the spatial patterns of development. Thus, after the Second World War, and confirming the then accepted conventional wisdom, both Caesar (1964) and Clark (1966) argued for the accessibility advantages of the London-Merseyside belt; Caesar envisaged the development of a 'megalopolis', analogous to that described by Gottmann in 1961, centred on Birmingham – the point of easiest access to the domestic market.

On that basis, one would not have expected there to be major development pressure in East Anglia, along the south coast and in the Bristol region, yet these have been notable for their growth during the last two or three decades. The so-called Cambridge phenomenon and the significance of the M4 corridor were both 'surprises'. Equally surprising was the post-war decline in manufacturing employment in Greater London (Dennis, 1978) and the recession which hit Birmingham in the early 1980s (Flynn and Taylor, 1986; Spencer *et al.*, 1986), neither of which events could be attributed to regional policy measures. (Note, however, that by the end of the 1980s industrial employment in the Midlands was again buoyant.) And some cities included in the northern part of the axial belt have experienced economic difficulties in the post-war period such that they are generally included in the depressed north of the country rather than the prosperous south – Merseyside-Manchester and Pennine cities such as Leeds, Bradford and Sheffield.

Early in the 1960s, some authors had questioned the logic of the argument that cumulative concentration *must* occur (e.g., Chisholm, 1964; Jewkes, 1960) but these doubts were generally brushed aside. Experience since then shows beyond any reasonable doubt that even though the south of England has prospered in relation to the north, the pattern of that prosperity was only imperfectly predicted by the prior application of cumulative growth models.

The United States provides a more dramatic contrast between prior expectations and the actual outcome. In his influential 1954 paper, Harris postulated the importance of market accessibility for industrial development and advanced the concept of 'market potential'. Using retail sales at the county level and generalized transport cost functions, he mapped the market potential for

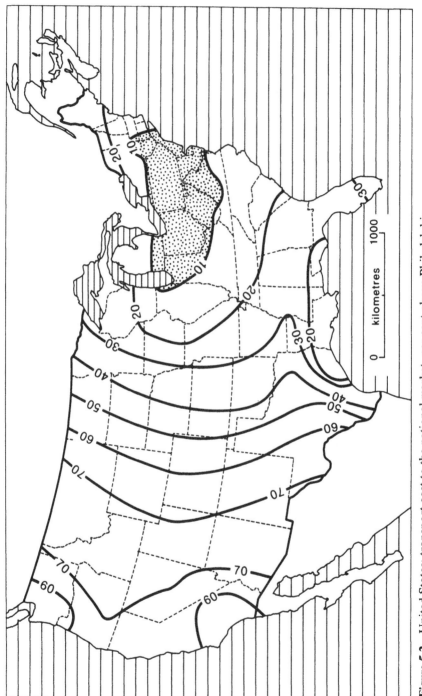

**Figure 5.2** United States: transport cost to the national market, per cent above Philadelphia
*Source:* Harris, 1954, p. 324

the whole of the United States (Figure 5.2). The stippled area, encompassing the area with transport costs less than 10 per cent above those of Philadelphia, approximates very closely to that which had one-half of all retail sales in 1948. Reviewing the period between 1914 and 1947, Harris noted that 60 per cent of the total growth in employment in non-local manufacturing had occurred in this region, which extends from New York to Illinois.

It seems fairly clear that Harris expected the northeast to maintain its dominance of the American economy, a view strongly supported by the cumulative causation model. Yet even in the 1950s this region was already ceding ground to the south and west (Moriarty, 1986; Perloff *et al.*, 1960). Although between 1955 and 1980, the manufacturing belt maintained its total industrial employment more or less constant, with only a marginal decline from 1970, the 'periphery' of the county experienced growth over the entire quarter century. By the late 1970s the 'periphery' had achieved an overall ascendancy (Agnew, 1987, p. 164). As Hansen put the matter: 'In recent years the once-vaunted heartland has experienced a reversal of fortune; it is now discussed and analysed primarily in terms of its problems rather than its successes' (1982, p. 208). Numerous scholars have examined the scale and causes of this considerable shift of regional fortunes, a shift which negates the kinds of prediction which had seemed reasonable in the mid-1950s on the basis of the cumulative causation model (e.g., Casetti, 1981; Fox, 1986; Perry and Watkins, 1977).

Two strands stand out in discussion of this dramatic shift in America's economic centre of gravity. The major population shift towards the west and south seems to have been at least as much cause as consequence of the shifts in employment. Rising standards of living and changing technology have made it possible for larger numbers of people to move to regions which, by virtue of climate or other characteristics, are preferred for living in (Chisholm, 1966, pp. 140–1). Employers respond, in that recruitment in these preferred areas is easier, especially for highly skilled workers who command premium wages. Thus, large numbers of people have quit the Snowbelt in favour of the Sunbelt. On the other hand, it seems that firms have been less prominent in moving, and that the regional differentials in employment growth are mainly attributable to differences in the birth and death rates of firms.

At the broad regional scale, experience in both the United States and United Kingdom suggests that the cumulative causation model is imperfect. The experience does not prove the theory to be wrong, only that an important ingredient is missing. For the present purpose, though, suffice to note that the confident assertions of the cumulative causation protagonists have not stood up too well in the face of reality.

A rather similar story is to be told of the normative version of cumulative causation, namely, growth centres. These are intended to establish self-sustaining growth in disadvantaged areas. Their success has been much less than had been expected. Friedmann and Weaver (1979, pp. 172–8) summarize the evolving view of growth centres in the regional planning process, from the first critical analysis published in 1971 to a remarkable clutch of studies published only five years later. Although some of the problems arose from inadequacies of plan implementation, a consensus emerged that planned growth centres had provided limited leverage for regional growth, in the centres themselves and also in the surrounding areas. (See also Dunford, 1988; Yuill, Allen and Hull, 1980.)

However, the rapid abandonment of growth centre strategies after little more than a decade of active promulgation was probably due to the over-optimistic expectations which had been engendered. In particular, regional planners had assumed that the development of the growth centres themselves, and their beneficial impact on the surrounding areas, would all take place much more quickly than in fact turned out to be the case. Drawing attention to the need for a time horizon of fifteen to twenty-five years, Richardson had this to say:

> Experience with growth poles is a textbook illustration of what can go wrong when regional planners act with insufficient understanding of the processes they are attempting to control, with unrealistic time horizons, with a cavalier approach to the difficulties of implementing growth poles, and with too rosy expectations of what a growth pole strategy can do for regional development.
>
> (Richardson, 1976, p. 8)

Put in another way, the short-term experience suggested that regional growth dynamics are far from 'mechanical', that input-output relationships cannot be assumed, and that therefore an approach which starts and ends by implanting some existing demand into a region and assumes that the desired supply relationships will materialize is, at the very least, inadequate. Much of the reason for this apparent lack of success for growth centre policies must lie, presumably, in the weakness of local multiplier and linkage effects.

## Fine tuning has its limits: motor manufacturing

Keynesian demand management policies were aimed at finetuning the economy. Some of the instruments were of a general nature but others were specific to particular industries. Thus, while the whole economy would benefit from success in managing demand, some sectors would experience particular problems of adjustment, as demand for their products oscillated sharply. If, as actually happened in the United Kingdom especially, the economy experienced bouts of general stop-go (Pollard, 1982), the adjustment problem facing all companies would be amplified in those industries used to regulate demand. Vehicle manufacturing was the major British industry to be thus affected. Being a substantial component of Britain's manufacturing economy, being initially the dominant (almost exclusive) supplier of the domestic market, and also being a major source of foreign revenue, the vehicle manufacturers were an obvious target for industry-specific fine-tuning measures. The consequence was not quite that which was intended:

> for twenty-five years, since the end of World War II, the government was able to use the UK motor industry as a policy instrument to help achieve a number of overall economic goals: increased exports in the 1940s to help

the balance of payments; regional relocation in the late 1950s to help areas of high unemployment; restrictions on income demand in the 1960s to help postpone devaluation. But in the 1970s the tables turned. Government policy as regards the motor industry had to be sharply modified as the government did what it could to help the much weakened UK motor industry fight off intense international competition.

(Dunnett, 1980, p. 121)

Of course, not all the blame lay with government, but there is little doubt that unexpected and frequent changes in government regulation had a seriously damaging effect (Pollard, 1983, p. 291). That by 1975 the industry was in a crisis so deep that it could have been terminal is hardly in doubt; three major enquiries into the industry were published that year, and the loss-making British Leyland Motor Corporation was effectively nationalized to become British Leyland (BL), at the start of what became a major rescue operation (Williams, 1983, pp. 217–18).

The parlous state of the British motor industry in the mid-1970s is demonstrated by the fact that, of fourteen major world companies listed by Dunnett, the four British companies lay among the five with the lowest value added per man in 1974, in the range £2,129 to £3,900, compared with General Motors (GM) (US) at £8,600 and Ford (US) at £7,966. A major contributory reason was the small amount of capital for each employee: the four British companies again were located among the bottom five (£920 to £2,657), compared with £5,602 for Ford (US) and £4,346 for GM (US). As for the return on capital, in 1977 this stood at 34 per cent for GM and a mere 1 per cent for Britain's BL (Dunnett, 1980, pp. 126, 134, 161). Why this state of affairs had come about is a complex story of management failure to manage the workforce, a belligerent and strike-prone tradition among the workers, poor design and marketing and poor control of quality. Not surprisingly, customers at home and abroad turned in increasing numbers to competing makes.

In this context, the frequent changes in hire purchase rules and the level of purchase tax plus VAT cannot have helped. Over the period from February 1952 to June 1977, when hire purchase restrictions were abolished, Dunnett lists 21 changes in either or both the minimum deposit and the maximum repayment period. Over the longer period from October 1940 to May 1979, there were 17 changes in the level of sales tax; of these changes, only two occurred in the same month as the hire purchase variations (Dunnett, 1980, pp. 88–9). Consistent forward planning for the domestic market was seriously jeopardized by these constant shifts, leading, *inter alia*, to a lack of innovation and lack of investment. And each time the restrictions were eased, output could not be raised immediately to meet the higher demand; consumers, fed up with long delivery times, turned to foreign suppliers and acquired a taste for their products. Especially in 1970–71, each 'go' phase saw a more than proportionate influx of imports.

The nadir was reached in 1981. Since then, import penetration has abated somewhat, and export performance has improved. However, the four 'indigenous' companies in the mass market now face competition from Japanese firms manufacturing in Britain and now taking a significant market share: the main plant is Nissan's at Washington, in the Northeast, but in 1989 both Toyota (Derby) and Honda (Swindon) announced major expansion plans. On any reasonable assessment, between the late 1970s and the mid-1980s, the industry achieved a significant renaissance, through the introduction of new models, new production methods, new management techniques and new ways of labour organization allied with a sharp reduction in the number employed (40 per cent in five years from 1978). Part of the manifest improvement must be attributed to the fact that in 1976 the government abandoned the habit of regulating the industry in attempts to fine-tune aggregate demand in the economy. The result is to be seen, *inter alia*, in markedly changed labour relations and management systems, aimed at providing flexible production with high levels of quality control (Marsden *et al.*, 1985). These changes have been facilitated by far-reaching modification of the legal position of trade unions since 1979.

Vehicle manufacturing was the industry most severely affected by regulatory policies designed to finetune aggregate demand. Similar effects were felt in other consumer durables liable to hire purchase controls, such as washing machines and refrigerators. And of course the secondary effects were felt in many industries, ranging from tyre manufacture and electrical components to sheet steel. Because industries do not have the instantaneous supply response which Keynesians assume, there is a limit to the control which a government can exercise, and it appears that in Britain that limit was substantially and consistently exceeded.

In their study of the West Midlands, Flynn and Taylor (1986) are in little doubt that the 'stop-go' period from the early 1950s proved to be very damaging for the regional economy, heavily dependent as it was on the industries used to control demand. On the other hand, Dow's (1964) general conclusion that macro-economic policy was destabilizing has been questioned (Cairncross, 1971). Matthews (1971) found some support for the complaints about 'stop-go' policies but concluded:

I suspect, however, that many of the complaints made about the effects of stop-go are not really about fluctuations. In part they are simply complaints about the average level of demand being kept too low; this applies notably to the complaints made by the motor industry.
(Matthews, 1971, p. 29)

The example of Britain's motor industry provides an illustration of the limits which exist in the power of governments to intervene and to shape the national and regional economy to their will. However effective such intervention may be in the short run, there are long-run costs with which to reckon (Brittan, 1983; Gatti, 1981). Equally important, the Keynesian assumption that supply-side issues are unproblematic is shown to be wrong. However important the short run may be, we ignore the long run at our peril.

Finally, Britain's motor industry also illustrates a truth which is manifest quite widely. When an industry is described as 'declining', as measured by employment and output, this should be treated as a description of current circumstances for that industry in the particular country. Britain's vehicle industry declined notwithstanding that domestic and international demand was increasing. The same has been true of several industries, of which shipbuilding is well documented (Thomas, 1983; Todd, 1985). Output from British yards has been in secular decline throughout the post-war period, despite rising world demand and production. Since British ships are, on average, about the same size as the worldwide average, which itself is quite small at not far off 10,000 gross tons, the decline cannot be attributed to the smallness of the estuaries, but must be due to a failure in adjustment, for whatever reason. The potential demand was there but this was not translated into orders and output. Both the motor industry and shipbuilding indicate that there have been serious failures at the industry level in providing customers with what they want. Under these circumstances, the troubles which have afflicted them – and other industries – must in part be troubles occasioned by failure to make the requisite adjustments, that is, supply responses.

## Conclusion

This brief examination of the Keynesian model of regional development in practice suggests that it provides, at best, an incomplete description of events, and policy prescriptions of limited utility. It would appear that the Keynesian model therefore shares certain characteristics with the neo-classical model. Neither provides a complete and convincing account of the way regional fortunes shift and change. Some would say, especially of the neo-classical approach, that its connection with reality is tenuous. But equally, the Keynesian version has not fared particularly well in accounting for the observed trajectories of regional growth, nor in providing adequate policy guidance as the basis for intervention.

Now it could be argued that either or both models were vitiated by the crisis which hit the advanced countries in 1973–4 and then deepened sharply between 1979 and 1983. Particularly of the Keynesian version, one might hold that it really did not have a chance. However, several of the strands of evidence we have reviewed relate to periods before 'crisis' became the current buzz word, or to dates sufficiently close to the onset of crisis that it is unlikely that radical changes in the economic system and behaviour of firms had in fact occurred. We can say with some confidence that it was not, nor now is, sufficient to wait for national economies to regain something approaching full employment for it then to be possible to resume a Keynesian strategy of regional intervention (Moore and Rhodes, 1982). The primary reason is that the available levers are now seen to be much less robust than had been thought, so that the effects of intervention are apt to be less than expected, and also unpredictable. The idea of redistributing demand within a national economy in order to achieve greater regional equality now seems to be less easily achieved than had been supposed, even when the circumstances are reasonably favourable.

Another source of disquiet for a 'pure' Keynesian was the realization that regional policy packages were having at least as much impact on the indigenous firms in assisted areas as upon inwardly mobile ones. However desirable such an impact might be, the cost in public expenditure was seen to be too great. Furthermore, stimulation of indigenous firms, whether by capital grants, provision of premises or by a subsidy on labour (the Regional Employment Premium in Britain) is difficult to distinguish from supply-side measures. To admit the need for such measures, however, is more than a little damaging to the coherence of the Keynesian position. To the extent that policy operated on indigenous firms and had the characteristics of supply-side measures, the logic of the Keynesian position was in question.

The period from 1973 to 74 which witnessed high and rising unemployment and accelerating inflation, accompanied by a widening of regional differentials, dealt a savage blow to both the neo-classical and the Keynesian models of regional development. That differentials appear to be continuing to widen, even though both inflation and unemployment have been brought to more reasonable levels, is not easily explained in neo-classical terms, except as an exogenous shock which has created a new disequilibrium situation which, in the fullness of time, will be rectified as the equilibrating forces get to work. In contrast, Keynesians would argue that the widening of regional differentials (unemployment and income) at a time of improving national prosperity reflects the reduction in the level of resources devoted to regional policy. On the evidence reviewed in this chapter, that argument is less than fully convincing.

Small wonder, then, that in the late 1970s and early 1980s, numerous scholars turned to other schools of thought, both in an attempt to understand what was going on and with a view to finding plausible prescriptions for policy. There was a fairly widespread sense of despair within both the neo-classical and the Keynesian schools of thought. Many were tempted to say with Mercutio, wounded in the defence of Romeo:

A plague o' both your houses!

# CHAPTER 6

# *Searching for a new way*

By the end of the 1970s the optimistic prospectus issued on behalf of the managed economy in the 1940s was widely regarded as false: unfulfilled and unfulfillable.

(Feinstein, 1983, p. 18)

The decade from the mid-1970s to the mid-1980s witnessed traumatic events in many advanced countries, as inflation was brought under control and rapid restructuring took place to cope with recession and the competitive threat posed by foreign manufacturers. Britain was, perhaps, the most severely affected nation, but other European nations were caught up with the same problems, and also the United States. The so-called 'British disease' received a lot of attention, and a string of publications appeared which were, in effect obituaries of the British manufacturing economy (e.g., Coates and Hillard (eds), 1986; Elbaum and Lazonick (eds), 1986; Gamble, 1981; Gardner, 1987; Weiner, 1981; Williams *et al.*, 1983). Similar concerns were expressed for the American economy (Bluestone and Harrison, 1982), and were reflected in the literature on regional development, since rapid restructuring at the national level implies sharp changes locally (e.g., Muegge and Stöhr (eds), 1987). Gertler (1986) has written about discontinuities in regional development, a theme which, in various guises, has informed much of the recent writing on both sides of the Atlantic, whether change is portrayed as the decline of monopoly capitalism (Graham *et al.*, 1988), the creation of new industrial spaces (Scott, 1988a, 1988b) or the wrecking of long-established regional manufacturing economies (Hudson, 1989). The sharpness of the changes, and the severity of the impact on individuals, is illustrated by the fact that in Great Britain, over the short period from 1979 to 1982, there was a net reduction of 1.6 million jobs. Confirmed redundancies, i.e., involuntary separation, amounted to 74 per cent of this total overall, but ranged from 56 per cent in the Southeast (the least affected region) to the daunting level of 90 per cent in Scotland (Martin, 1984, p. 451). Over the slightly longer period from 1977 to 1983, 16 firms which collectively employed 1.259 million people in the earlier year had only 0.728 million on their payrolls six years later (Martin, 1986, p. 11).

That these changes were drastic and, from the viewpoint of previous decades, unexpected, is not in doubt. On the other hand, there is an unresolved debate as to whether the severity of the recession could have been mitigated if governments in both the United Kingdom and the United States in particular had pursued less rigorously monetarist policies than they initially did. In the present context, however, the issue of main interest is the following. Given that both the neo-classical and the Keynesian approaches to economic management stood largely discredited, and given that the monetarist medicine was extremely unpalatable (and, some believed, in any case heading for disaster), could some other view of aggregate growth and of regional development be found? It was natural to ask whether a body of thought existed but which had been generally overlooked, to which one could turn for guidance.

Three such bodies of thought lay to hand, one of which we have explicitly discussed in Chapter 3, while the other two are implicitly acknowledged in Chapter 2. For some observers, the radical or neo-Marxist school of economic thought seemed to offer both a perspective on economic changes and policy prescriptions relevant for regional affairs. In addition, with acceptance of the role of innovation in the development process generally, the idea of long waves of economic prosperity and recession gained credence, not least because the 1970s to 1980s downturn had apparently been predicted. Associated with the long-wave thesis is the idea of profit cycles in manufacturing industries. In this chapter we will explore these ideas to see what light they throw on the problem posed by Feinstein at the beginning of this chapter.

## Radical approaches

Marxist thinking emphasizes the contradictions postulated to be inherent in capitalist production, and in particular the idea of recurrent crises as 'overaccumulation' occurs and as the 'inevitable' decline in profitability of private enterprise takes place. Such crises throw into sharp relief the conflicts between the workers on the one hand and the capitalist class on the other. Given the assumption that profits are inversely related to the level of wages, the owners of capital will shift investments away from high-wage toward low-wage regions, either within a country or abroad, causing unemployment in the formerly prosperous regions to rise. If investment is shifted to low-wage developing countries, then all the regions of the advanced nation would suffer. During a recession, these conflicts of interest between workers and capitalists will become particularly acute, with the capitalist class often cast as the villain of the piece as workers struggle to protect their jobs, while capital, trying to reverse the decline in profits, seeks to shed labour and rationalize operations. Such conflict is manifest at the regional scale, focusing around the threatened closure or slimming down of particular industries or plants. To the extent that the capitalists win this battle, so do they recreate a 'reserve army of labour' (i.e., unemployment) which provides the conditions necessary for a period of renewed profitability (Harvey, 1982).

Numerous bitter industrial disputes marked the late 1970s and early 1980s in several countries, and appeared to support this interpretation. The steelworkers of Lorraine, the coal miners in Britain and workers in America's mid-west motor industry all sought to confront the threat of closure and redundancy, in particular by means of strike action. It was manifest, too, that decisions regarding rationalization plans, and the consequential geographical patterns of redundancy and closure, were taken at head offices located far from the scene of immediate action. In their fight for survival, firms will cut costs (by shedding labour), and will engage in strategies of merger and rationalization, apparently without reference to their employees. In face of employment contraction, workers have no control over events, nor even a voice to be heard. Thus, a region facing recession does so, it is argued, on account of decisions taken by the owners of capital. On this analysis, regional problems are problems experienced by regions but for which the region itself, and its inhabitants, is not responsible (Anderson *et al.*, 1983; Markusen, 1987; Massey, 1979; Massey and Meegan, 1978; Scott and Storper (eds), 1986).

This reading of events further suggests that capital, faced with the inevitability of crisis, will seek to avoid a grand crisis which afflicts the whole of a nation simultaneously. Much better to have a series of local crises, each of which can be contained without jeopardizing the whole fabric of capitalism. Thus may one paint a picture in which no region is exempt from periodic savage recession, which will leave, if you will, a local economic wasteland, ready for the next cycle of accumulation. As one noted Marxist scholar has put it: 'The geography of uneven development helps convert the crisis tendencies of capitalism into compensating regional configurations of rapid accumulation and devaluation' (Harvey, 1982, p. 428). Since, in the early 1980s, recession had hit not just the traditionally less prosperous regions, but also regions which had hitherto enjoyed long prosperity, Harvey's analysis enjoyed the support of actual experience.

Three rather different conclusions may be drawn from the situation outlined, of which the most extreme is the call for a socialist revolution, nothing less than putting an end to the capitalist system. This idea can be found in the regional literature, stated quite explicitly:

> parochial solutions to regional problems . . . are not solutions at all. This is not to say that nothing can be done at the local level, but that most of what can be done will have to be in alliance with other areas as part of a wider strategy to wrest economic power from the divisive control of capital.
> (Morgan and Sayer, 1983, p. 46; see also Richardson, 1984)

While such a revolution would end exploitation as formally defined in Marxist thought, it would leave unresolved the problem of what happens if one group of workers (or one region) succeeds in gaining a competitive edge by virtue of price, quality of product or novelty of the articles produced. As Morgan and Sayer recognize: 'other regions would become less competitive' (p. 44). Given the interdependence of the world economy, and given the fact of continuing technical innovation, a change in the ownership of the means of production does not directly address the underlying problems. In any case, however attractive socialist revolution might have appeared in the early 1980s, events in the USSR and Eastern Europe since the advent of Mr Gorbachev suggest very clearly that all is not well in the lands of socialism, just as the 1989 popular uprising in China was a cry for freedom from the 'socialist' tyranny.

A less extreme conclusion is the proposition that, since it is capital which causes the problems, capital should be controlled. This view was clearly spelled out by Holland (1976a and 1976b), in a two-volume examination of regional matters, one volume being given the title *Capital versus the Regions*. Holland's central thesis is that national and regional fortunes are determined by what he calls the meso-economic sector – the very large companies which control a large portion of economic activity and are otherwise known as multinationals. He cites, for example, the 100 firms which controlled about one-half of Britain's manufacturing output in 1970 (Holland, 1976a, p. 138). Given the relative immobility of labour and the facility with which the meso-economic sector is assumed to be able to switch its investments, the kernel of the argument is as follows: 'the key to unlocking the underlying imbalance between the regional distribution of capital and labour lies in the control of capital' (Holland, 1976b, p. 136). On that view, the solution is for governments to enter into agreements with the large firms concerning the nature and location of their investments, or for governments to take direct control by nationalization or other means. To the extent that planning agreements are used, Holland envisaged that this would have to be done at the level of the European Community, since otherwise the big firms would play one country against another. Even then, if the conditions imposed by the Community were deemed to be too onerous, presumably the large firms would choose to invest elsewhere in the world. An alternative is the outright nationalization of the one or more industries which are dominant in the less prosperous regions, as advocated by Damette (1980) for the Lorraine steel industry. This industry was undergoing major retrenchment at that time.

Holland's two books were published in the very year that the British government abandoned the idea of detailed intervention in individual industries to manage aggregate demand. His proposition also has to face the costs which intervention, in the form of planning agreements, would impose. Such costs set limits to what can realistically be achieved without jeopardizing the long-term prosperity of the industries (see pp. 98–101). As for nationalization, it is of some interest that Hudson (1986, 1989), writing of northeast England, regards the creation of what he describes as an 'industrial wasteland' as being due in considerable measure to the rationalization process of nationalized industries – coal and steel – as they sought to convert unprofitable operations into profitable ones. Whether an industry is nationalized or not, it cannot escape the pressures exerted by competitors at home and abroad. Consequently, nationalization, or other forms of public control, is not a policy which offers a long-term solution. Indeed, since 1979 there has been a widespread perception that state-owned enterprises have lacked sufficient commercial drive and are consequently less efficient than is necessary to face the competitive pressures at home and abroad. As a result, in much of the western world state-owned industries have been shifted into the private sector by various measures of privatization.

The third solution which can be traced to the radical analysis of capitalism is that disadvantaged regions would benefit by partial or complete 'closure', i.e., by severing economic links with other regions and becoming largely self-sufficient. Thereby, it is believed, the full rigours of capitalism could be kept at bay. The main inspiration for this idea, whether explicitly acknowledged or not, lies in the underdevelopment school of thought, and especially in the idea of unequal exchange.

'Uneven development', the regional version of the underdevelopment school of thought, is an imprecise concept:

> Uneven development is not simply a geographical pattern but a process rooted in the capital-labour relation, in the division of society into classes, in the uneven relation between use-value and exchange-value, in the anarchy of the social division of labour, in the inequality of the relative mobilities of capital and labour, in the unequal gains and losses from competition, and in the increasing scale of production.
>
> (Morgan and Sayer, 1983, p. 45)

A similar view has been expressed by N. Smith (1984, p. xiii), uneven development being: 'the geographical expression of the more fundamental contradiction between use value and exchange value', giving rise to 'development at one pole and underdevelopment at the other'.

If this thesis is accepted, the policy conclusion is clear – sever the economic links and thereby stem the outward flow of resources from the disadvantaged region(s) arising from unequal exchange. 'Territorial closure' is the term which Friedmann and Weaver (1979) used, comparable to the idea of 'de-linking' advocated by some at the international level (Díaz-Alejandro, 1978). Friedmann and Weaver do not suggest that regions should become completely self-sufficient, but argue for selective restriction of trade across regional boundaries. The idea achieved considerable prominence in 1981 with the publication of *Development from Above or Below?*, edited by Stöhr and Taylor, and Weaver's 1984 book on regional development.

The aim of development from below would be to serve the needs of the regional (local) population, to ensure that the basic necessities of life are provided and that any surplus which accumulates is devoted to enhancing the welfare of those people, not some remote corporate entity in another region or nation. Some form of collective or socialist organization would be needed. Control would have to be exercised over trade in commodities and services across the regional boundary, as well as over the movement of capital. This need for control over transactions across the regional boundary is a fatal weakness in the idea of regional development from below, in that no region in the sense we are using the term (see pp. 3–5) actually has the power to exercise such control; were it to do so, it would look much more like a sovereign nation than a regional division of a larger state.

The argument for partial system closure can be derived from a different intellectual tradition. International trade theory has for long recognized the existence of scale economies in production. As it will normally take several years for a new plant's output to build up sufficiently for the main scale economies to be obtained, and since in that period an industry is vulnerable to imports, the case has often been put and accepted that temporary protection in the form of import tariffs or quotas is justified. Once output has achieved the relevant threshold, protection can be withdrawn. However, the power to provide infant industry protection lies with the national government, not with regional administrations (other than through their procurement policies).

The idea that development from below can take place on the initiative of a region and in the absence of co-operation from central government is clearly vain. On the other hand, the idea that local initiative could be important, that local action can influence the course of events, is an idea of great significance. It is, of course, an idea that has a respectable history in the manner in which individual entrepreneurs and civic leaders have played leading roles in local development. If we can set aside the flawed doctrine of unequal exchange, then the idea of development from below is an assertion that however important external forces may be, a region is not an entirely passive reactor to events over which it has no control. But, as we shall develop more fully in Chapter 8, a region must act in concert with the central government.

Socialist revolution and control of the meso-economic sector, however attractive they may appear to be at first glance, provide no lasting solution. System closure is not in fact an option available to a region, even if the argument for such a course of action were persuasive (which it is not). On the positive side, it is clear that some radical thinkers are searching for a basis on which to assert that regions should have some control over their own destinies.

## Long waves of development

It is a well-established fact that nations experience cyclical fluctuations in their economic fortunes, and the literature on the subject is positively vast. The longest cycle to be identified, with a periodicity of about fifty years, was reported early in the twentieth century (van Duijn, 1983), but attracted little attention until Kondratieff published his own studies, first in German and then, almost a decade later, in English (Kondratieff, 1935). Whether these long waves really do exist is still a matter of controversy (see Solomon, 1988). Kondratieff's work strongly influenced Schumpeter (1939), but thereafter the whole genre of studies concerned with cyclical phenomena temporarily fell out of fashion with the rise of the Keynesian orthodoxy. Keynes offered the prospect that severe cyclical oscillations would be a thing of the past, relevant for the history books only, since henceforth governments would be able to control demand to maintain full employment. Throughout the 1950s and the 1960s, only a limited amount of work was published on long waves (see Kleinknecht, 1987), and virtually all of that was outwith the mainstream of economic thinking and the analysis of regional growth processes.

This all changed with dramatic speed in the 1970s, for a reason that is readily apparent:

Neither Kondratieff nor Schumpeter embarked on speculation about the future. But if they had, then the operation of the cycle would have predicted a steady expansion of the world economy from about 1945 onward – and then a descent into depression from the 1970s. It is small wonder that some people are starting to take Kondratieff seriously again.

(Hall, 1981, p. 535)

Was it a coincidence that the downturn of the 1970s and early 1980s was thus 'predicted' or was there a firm basis to the long wave theorizing? If there was a firm basis, what does this tell us regarding aggregate growth performance and the fortunes of individual regions? To answer these questions, we must first look at the basis for long wave theorizing.

Long wave analysis began with the empirical observation of periodicities in price series data. Similar, but less marked, periodicities have been observed in output data (Cleary and Hobbs, 1984). For present purposes, therefore, the practical starting point remains the evidence obtained from price data, showing periods of relatively rising prices (associated with prosperity) followed by relative decline (periods of recession). Dating the turning points (the peaks and troughs) is a problem which has attracted a lot of attention, despite which there is not yet a clear consensus (Kleinknecht, 1987). Our initial problem, therefore, is that long wave analyses are based primarily on price data, not output data, and that the dates of the turning points remain to be resolved. Consequently, there is still considerable doubt concerning the nature of the empirical phenomenon for which a theoretical explanation is to be sought. Beyond that, however, is the fact that thus far only five cycles have been identified – four complete ones and the fifth which is currently said to be occurring. With such a small number of cycles available for analysis, formidable problems are raised concerning the reliance that can be placed on the empirical patterns observed and the causal mechanisms which may be suggested.

In fact, several mechanisms have been offered to explain the transitions from upswing to downswing, and then the subsequent upturn to renewed faster growth. These mechanisms include the recurrence of scarcities in the supply of primary products; the decline in profitability of business; periodic investment booms to replace obsolete infrastructure, such as transport systems and sewers; and the temporal bunching of innovations. Numerous scholars have described these competing theories; none has provided a satisfactory basis for selecting one in preference to another (e.g., Freeman, 1983; M. Marshall, 1987), or in providing a fully satisfactory synthetic theory (Goldstein, 1988).

The explanation which currently receives the most attention is the last of those just listed, the bunching of technological innovation. This was first propounded by Schumpeter (1939) and the idea was given considerable impetus with the appearance forty years later of Mensch's book *Stalemate in Technology. Innovations overcome the depression*. The thesis has two elements, concerning the rate at which inventions occur and the rate at which these are converted into practical applications (innovations). It seems necessary to postulate that during periods of prosperity, many inventions remain unused. The resulting accumulation will only be exploited when serious recession occurs and firms, desperate to find ways of restoring their fortunes, convert the accumulated inventions into innovations – of product and/or process.

Much judgement is required to select the innovations which will be included in any tally which is designed to test this thesis. Even greater judgement is necessary if the relative importance of innovations is to be assessed. For example, how is one to rank the importance of the Bessemer technique for steel making, the introduction of electricity and the role of the internal combustion engine? Finally, the criteria by which an innovation can be given a date are imprecise, so that the date assigned is also a matter of judgement. The empirical evidence for the bunching of innovations is, therefore, somewhat questionable. Equally in doubt is the process which might lead to inventions remaining unexploited until a period of major recession occurs.

Notwithstanding these doubts, long wave analyses have entered the literature on regional growth and development. In essence, the argument runs as follows. During an upswing phase of prosperity, technology will remain reasonably stable, reflecting the relative scarcity of innovations. Consequently, there will be a fairly constant pattern of growing and declining industries, the location requirements of which will remain broadly invariant over time. However, when the next downswing occurs, many long-established industries will find themselves in real trouble. At the same time, the bunching of innovations causes new products to come on the market, which will either by manufactured in the 'old' industries, as they adapt their product range, or in entirely new companies spawned by the innovators.

If we treat long wave theory as a *descriptive* framework, there can be no quarrel with the following proposition. The innovation process does not operate uniformly through time, with the result that there are periods when the sectoral composition of production changes rather rapidly, and others when the modifications are less dramatic. This is, after all, the form in which Rostow (1978) has written of the leading sectors of development. Whether these fluctuations in the tempo of innovation occur in a regularly recurring cycle of more or less constant duration seems still to be an open question.

If we wish to translate this temporal description of aggregate change into the resultant geographical patterns, some important analytical steps must be taken. First, we need to know whether the new industries have locational needs that differ from those of the older industries. Second, we need to know where these needs can best be satisfied. These two questions are the stuff of traditional industrial location analysis. Unless one can supply evidence on these questions in an *ex ante* manner, one is likely to be driven to make inferences from the observed patterns of old and new locations, with the inherent danger of engaging in an argument which is circular in nature.

To date, the main use of long wave theory in regional analysis has been as a descriptive framework of a suggestive kind. As it stands, long wave analysis gives no direct guidance on whether a particular new industry will develop in one kind of location or another, or to what extent it will locate in both 'old' and 'new' industrial regions. Thus, no direct guidance is given concerning the fortunes of particular regions. (Hall, 1985; Hall *et al.*, 1987; M. Marshall, 1987; Markusen *et al.*, 1986; Scott, 1988a, 1988b.)

It is common knowledge that in the last two decades, or thereabouts, there have been some dramatic examples of new industrial regions emerging, associated with modern 'high tech' products. The San Francisco area in California

('Silicon Valley'), Orange County in Southern California and the Cambridge region in England are well-known examples, as is the remarkable growth of new industries in the southwest of the United States and along Britain's M4 corridor. On the other hand, Britain's Birmingham has ridden out the four full cycles identified by long wave theorists, and both Manchester-Lancashire and Scotland have notable concentrations of modern industries associated with the present (fifth phase); and while it may be an exaggeration to describe Massachusetts' industrial revival as a 'miracle', the transformation of old industrial areas in that state and elsewhere in New England has been remarkable (Markusen *et al.*, 1986, p. 173).

There can be no doubt that innovations in the microelectronic, computing and biotechnology fields in particular have, in recent years, spawned major product and process innovations, which may be a foretaste of even more dramatic changes in store. Equally, there is no doubt that important new industrial areas have emerged. But it is equally true that many older industrial regions have made partial or even highly successful adjustments to the emergent new industries.

In spatial terms, therefore, the real question which long wave theory poses, but cannot itself answer, is the following: Given that innovations generate new industries, what are their location requirements, how do these compare with the needs of older industries and where may these new needs be most efficiently met?

Long wave analysis introduces the supply-side in three different senses. First, whereas it is common to visualize technological development as a continuous process, the postulated bunching of innovations indicates that there are important discontinuities in the system. Why and how that bunching may occur involves questions about both the supply of inventions and the trigger mechanisms which convert inventions into innovations. Second, the mechanism which is suggested whereby the downswing is converted into an upswing is essentially that new products generate new demands – a version of Say's Law. Thus it is that long wave theory at the aggregate level clearly raises supply-side issues which had hitherto been accorded rather little attention. The third supply-side aspect relates to the geography of the new industries. Why do they locate sometimes in 'new', sometimes in 'old', industrial areas? What are the regional characteristics which make for success in attracting the new enterprises, and to what extent may these characteristics depend on natural resource endowment and location, or how far do they depend on man-made attributes? To the extent that it is the man-made features of a regional economy which are important, how far may these be consciously modified? That question is no more than a reformulation of a perceptive observation made by Thompson as long ago as 1965. Aware of the work of Kondratieff, he commented: 'We might postulate a crisis theory of human behavior in regional economic development: a community rises to the occasion in a variation on the Toynbee theme of "challenge and response"' (Thompson, 1965, p. 18).

## Profit cycles

In our review of trade theory, reference was made to the product cycle as one manifestation of the technological leads and lags which give rise to international trade. This idea can be applied to patterns of regional development, whether or not innovations (and hence product cycles) are bunched in time, as suggested by long wave theory (Steiner, 1985). The product cycle can be generalized to the idea of profit cycles (Hall, 1985, p. 2), a suggestion taken up and examined at length by Markusen (1985). She postulates that industries experience a cycle of profitability, which moves from zero to very high and then back to zero. The argument runs as follows. A new product is invented and an industry grows to produce and market it. Initially, development costs absorb all the available revenue, with the consequence that profits are negligible, even negative. Once the initial teething problems have been sorted out, Markusen envisages that there will for a time be only a limited number of producers. On the assumption that the product is in considerable demand, the firms engaged in its production will enjoy a period of super profits; as oligopolists, they can control the market in substantial measure and reap substantial benefits thereby (see pp. 33–4). However, the existence of these unusually high profits will attract more firms into the business, so that total output will rise relative to the market demand, and profits will be driven downward to the normal level associated with a competitive industry. Thereafter, the industry will steadily become obsolete and profits will decline towards zero. When this point is reached, the industry will either close, or move to another location (abroad?) where profitable production can be maintained.

If individual industries experience the profit cycle which Markusen postulates, then one may envisage a fairly direct connection with the processes of regional growth and decline. At the early stages, with a limited number of firms, the regions in which they are located will experience a high level of prosperity. With the entry of additional producers and the beginning of a squeeze on profits, both the new entrants and the longer established firms will have to consider their choice of location carefully. Attempts to safeguard profits may tend to the abandonment of the initial production sites and the development of others where production costs are lower (usually on account of lower wage costs). Markusen envisages that even this strategy cannot postpone for ever the demise of the industry.

Having developed the profit cycle concept, Markusen reviews several important sectors of manufacturing in the United States, and the experience of many communities. On the basis of that evidence, she comes to the following conclusion: 'The point is that the rise and decline of a region's economic base are predictable developmental events' (Markusen, 1985, p. 279).

Presented in the above terms, the profit cycle suffers from a number of weaknesses. First, it seems to be assumed that a new product necessarily implies the creation of a new industry, as was clearly the case with modern steel and motor vehicles. However, many new products are manufactured by existing companies which alter their product range, as has happened over a wide range of chemical and pharmaceutical products. Second, Markusen concentrates on product innovations and pays little attention to the process innovations which may have a major impact on the way manufacturing is carried out and consequential effects upon the level of profitability.

The final problem concerns the ability of a region to attract new firms at an early stage in their development. Markusen takes it as axiomatic that large, oligopolistic firms create conditions in their immediate environs which are unsuitable for new enterprises. Consequently, she visualizes a continuing process whereby regional boom is necessarily followed by bust and abandonment. Although she does allow the possibility that the adverse stages of the profit cycle may be postponed by reorganization of the industry in question, or that the decline stage may be compensated by the attraction of new firms, little weight is attached to either possibility:

> The developmental implication of the profit cycle model is that regions with largely mature sectors will be able to sustain their local economies only by continually renewing the economic base with the addition of new sectors. There are many historical instances of such success: New Bedford switching from whaling to textiles, Minneapolis from a timber economy to flour milling, to computers, New York from apparel and cigar making to a center of financial exports. But they are increasingly less common. Because world capital markets have become so integrated, the profits from a region's local economy are no longer necessarily reinvested within the region, as was the case in the nineteenth century. The opening up of alternative sites in the Third World makes it easier for profit takers to escape the older economies that are dominated by corporate leadership and working-class cultures antagonistic to new sectors. And the evolution of the large oligopolistic corporation and of institutionalized capital-labor conflict tends to make older regions increasingly unattractive to the badly needed newer sectors.
>
> (Markusen, 1985, p. 289)

This is a pessimistic conclusion which implies that there is no hope for a region which is dominated by a large oligopolistic industry when that industry contracts or closes. Yet there are recent examples where in fact great strides have been taken to replace such jobs with new ones. Pittsburgh (United States) and Corby (United Kingdom) were both steel towns which in the post-war period have successfully adapted to the closure of their plants. Sheffield has embarked on ambitious plans to the same end, seeking to emulate the remarkable turn-around which has been achieved by Glasgow (Cheshire, 1990) in the last fifteen years or so.

It clearly is the case that industries do experience temporal variation in their profitability, and it is equally clear that associated with these changes there are changes both in total employment and in location. However, to postulate an inevitable progression, in the form of a cycle, which leads to closure of plants and their relocation abroad denies the scope for resurgence in those industries. In recent years, for example, the steel industries of both the United Kingdom and the United States have fought back from very adverse situations, to create efficient and competitive industries which now look capable of surviving for the foreseeable future. The British steel industry incurred losses of £1,784 million in 1979–80; by 1988–9, it returned a profit of £593 million (Keeble, 1990a). The motor industry is another that has experienced considerable improvement in its profitability in old locations, as well as the establishment of new 'green field' plants. Investment in modern production methods and reorganization of work practices have enabled profits to rise again, giving hope that these industries will in fact postpone (indefinitely?) the postulated terminal decline in profitability. Meantime, other industries which had abandoned some more advanced countries for cheaper production sites in the Third World are now returning (Milne, 1989).

We are thus led to much the same conclusion that we reached in our consideration of long wave theory. Manifestly, industries do experience temporal variations in their level of profitability. However, it is not persuasive to suggest that there is an inevitable cycle which gives rise to predictable outcomes in terms of regional development. The important question for us to consider is the nature of the links between an industry's performance and the region(s) in which it is located, and the role that regional initiative may play in fostering that industry or in attracting replacements.

## Conclusion

The three sets of ideas reviewed in this chapter share some common themes. In their various ways, they can all be thought of as attempts to find some pattern and order in the circumstances of the world which clearly no longer conform to the verities that had hitherto been widely accepted. To search for order and pattern, and hence to look for the basis on which clear policy can be formulated, is the essence of the scientific endeavour. However, for the reasons indicated in each case, the three bodies of ideas do not measure up to this aspiration.

In all three cases, the question is raised concerning what, if anything, can be done locally (but probably with the help of central government) to facilitate change and adaptation. At one extreme, localities and regions are regarded as the playthings of external forces (the decisions of capitalists) over which they have no control. Such a view has some similarities with the Keynesian export multiplier, in the sense that this too postulates exogenous impulses to which the region responds in a passive way. On the other hand, the 'development from below' concept challenges this deterministic, and essentially pessimistic, view of the mechanisms of regional growth, posing the question of what it is practical for a region to do in an active manner to help itself. That theme crops up again when we examine both long wave theory and the idea of profit cycles. In both cases, a mechanical, deterministic application of some 'law' is not a viable approach. Why it is that some regions adapt and make good homes for new industries, while others do not, cannot be answered directly. Similarly, why some industries in some localities succeed in postponing or reversing the onset of diminishing profits clearly depends on some interplay of factors specific to the industry and to the region. Consequently, we may view the problem in the manner envisaged by Thompson, as a matter of challenge and response analogous to Toynbee's approach to the rise and fall of civilizations.

An important dimension, therefore, of the dynamics of regional growth and decline is the intersection of the interests and responses of entrepreneurs and firms on the one hand, and the interests and responses of workers and inhabitants in the regions on the other. To the extent that the actions and reactions of the citizens may affect the decisions of businesses, patterns of regional growth will be modified. This is another way of saying that the nature of the adjustment processes will have an impact on growth and on prosperity. In effect, the adjustment process involves supply-side issues, which thus far we have not addressed in detail. Is it the case that systematic incorporation of supply-side issues could go some way to resolving the apparent dilemma with which Chapter 5 ended and which clearly has not been resolved by the material reviewed in this chapter?

# CHAPTER 7

# *Taking the supply side seriously*

> An adequate level of aggregate demand is the *necessary* condition to full
> employment ... the supply side adds the *sufficient* condition.
>
> (W. R. Thompson, 1965, pp. 207–8)

Supply-side matters have been touched upon at a number of points in the
preceding discussion though neither of the two main schools of thought which
we have discussed gives much attention to them. For neo-classicists, market
imperfections (and hence supply-side problems) are assumed away, or treated
as exogenous variables – namely, population growth and technical change. By
definition, supply-side problems do not exist in this framework of thought. To
the extent that neo-classical theory is modified to accommodate the manifest
fact that disequilibrium conditions do exist, it is generally assumed that supplies
of capital and labour will adjust over time, although this is recognized to be a
long-term rather than a quick process. Keynesians, giving demand the pre-
eminent place, are apt to assume that supply adjustments are not problematic,
at least in situations where unemployment exists above the level deemed to be
full employment. This confidence in supply adjustments was expressed by the
British Treasury as recently as 1985, in evidence to a Select Committee on
Overseas Trade. At a time of considerable concern about deindustrialization,
when British manufacturing had experienced severe retrenchment and the
balance of commodity trade looked unfavourable, especially if oil revenues
were excluded, it was a matter of considerable interest what would happen
when oil resources began to be depleted. The Treasury argued that when the oil
began to run out, there would be an *automatic* resurgence of industrial output.
The Select Committee was less sanguine.

Deeply embedded in both the neo-classical and the Keynesian thought
systems is the idea that the factors of production, and also the products of the
production process, are homogeneous. Radical/Marxist thinkers are apt to
make a similar assumption. Not so in the theory of monopolistic (imperfect)
competition, in which the differentiation of products is emphasized, as also the
existence of spatial monopolies. The imperfection of markets implies that
adjustment is problematic, and that supply rigidities may be important.
Furthermore, it can readily be shown that under monopolistic competition
there is apt to be a divergence between the interests of the monopolists and of
the consumers, such that regulatory intervention is essential if competition is to
be maintained and consumer interests protected. Monopolistic competition

theory leads to the apparently paradoxical situation that 'free' markets can only be maintained by regulation, in the manner of an umpire enforcing rules of play. Implicit in this approach, therefore, is the idea that the supply side of the economy is in fact problematic and that steps may have to be taken to ensure that factor and product markets do respond adequately to demand. From this viewpoint, it is immediately evident that concern with supply-side issues is in no way to be equated with a neo-classical, *laissez-faire* approach to the economy. The starting point is that supply responses are imperfect and the basic question is: What steps are needed to achieve what level of perfection in the relevant markets? This puts the spotlight on the whole range of supply bottlenecks, which the 'hidden hand' of market forces cannot be assumed to deal with. Supply-side thinking, therefore, stresses the role of incentives (market signals) in assisting the efficient allocation of resources, which is a neo-classical view, but emphasizes the necessity for continuing intervention to achieve this objective. Such intervention is primarily that of the regulator or umpire, to limit the power of monopolies and to ensure a reasonably competitive economy, without necessarily implying that the regulator best knows how resources should be allocated. The position is, therefore, somewhere between the two extremes noted by Brittan: 'While the basic mistake of the interventionists is to exaggerate the effects of government action, the mistake of the free-market school is to exaggerate the benefits of government withdrawal' (1983, p. 13).

Just as supply-side thinking is distinct from neo-classical and Keynesian thought, so also is it distinct from monetarism, even though many writers in the 1970s and early 1980s regarded them as synonymous. Monetarism stresses the role of money supply in the management of an economy and in that sense is part of the supply-side critique of Keynesianism. However, pure monetarism has been out of fashion since the mid-1980s (Dow and Saville, 1988), for at least two reasons. First, the price of money is not solely determined by the quantity that is in circulation, or even by the quantity multiplied by the velocity with which it circulates. The demand for money also has a part to play. Second, the definition of 'money' is not easy. Money is not just cash in hand and money at the bank; it is also the deposits held with building societies, credit that may be obtained (with or without security), Eurobonds and many other things. To control the multiple aspects of money supply in a modern economy is no mean task, and indeed is held by some to be impossible in any absolute sense. The fact that pure monetarism has ceased to be in vogue does not prove that control over money supply is irrelevant, only that it is difficult in practice. As the dogma has been softened, we can see what was previously obscured – that monetarism is but a single aspect, albeit an important one, of a wider concern with the hitherto neglected problems of supply in the economy. In this sense, therefore, supply-side thinking is not an alternative to Keynesian analysis but an essential complement.

In any case, 'demand' does not exist as an objective given, waiting to be supplied. When a business invests to produce goods or services, it is doing so in order to meet *expected* demand in the future. These expectations may or may not be realized. Investment, therefore, may be regarded as a supply-side act, designed to meet expected demand either for existing goods and services, or for

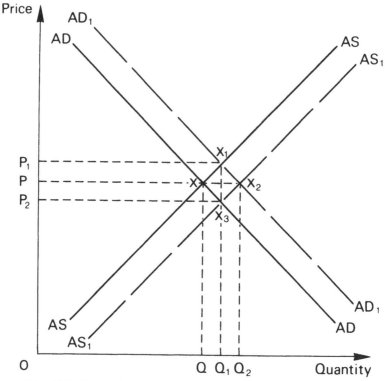

**Figure 7.1**   Shifts in supply and demand schedules

new ones not previously marketed (Gilder, 1981). The quality of that investment, and the cost and quality of the resulting output, may well have a bearing
on the scale of demand which is realized in practice, whether this is above or
below expectations. Therefore, to the extent that there are links between
supply, demand and price (see Figures 3.1 and 7.1), it is implausible to suggest
that all the adjustment occurs on either the demand side or on the supply side:
'Intermediate positions are possible, and more appealing, in which interaction
is allowed for between forces on the supply side and forces on the demand side'
(Matthews *et al.*, 1982, p. 17). The interaction may well be reflexive in character,
so that the supply and demand schedules are not in fact totally independent of
each other, even if it is convenient to retain the fiction that they are. In the
present context, however, the point to emphasize is this: when we consider the
real world, it is implausible to assume that problems arise on only *one* of the
two sides of the supply–demand relationship; both need to be considered.

Obvious though this point may seem, macro-economic thought has for a long
time neglected the supply side. Evidence for this statement may be found in the
first eleven editions of Samuelson's influential text, *Economics*, which was first
published in 1948. Discussion of macro-economics focused almost exclusively
on demand issues. Only with the appearance of the twelfth edition (Samuelson

and Nordhaus, 1985) has the discussion become more balanced, with a consideration of both supply and demand responses. Thus, it is only from the mid-1980s that one can confidently say that the significance of supply-side matters has been fully recognized in conventional mainstream macro-economic analysis, as the new 'authorized version' of economic doctrine (*Economist*, 14 March 1987, p. 93).

Keynesian thinking emphasizes the short run, to the virtual exclusion of long-term considerations. In contrast, neo-classical thinking about the real world, as distinct from its theorizing about how economies should work, stresses the long-run nature of adjustment processes. Supply-side thinking also emphasizes the importance of the long run. As W. R. Thompson put the matter: 'We might generalize to the effect that the longer the time period under consideration, the greater the relative importance of supply – local resource endowment and industrial culture' (1965, p. 37). However, the emphasis now lies with the capacity of a regional economy to adapt, to absorb innovations and to be the focus of entrepreneurship. To take a long-term view implies abandoning the assumption that there is a constant relationship between economic variables. Technological change, changes in personal income and consumer preferences, the emergence of new sources of competition and changes in regulatory and fiscal arrangements all have an impact on the market relationships. If we abandon the myopic concern only with the short run, and take an interest in the long run as well, then it becomes immediately apparent that we can no longer assume the fixity of relationships between variables. Instead, account must be taken of changes in consumer preferences, the emergence of new products and processes, changes in the competitive environment, etc. Thereby, short-term verities dissolve into probabilities or even just possibilities, with the result that:

It is one thing to say that the world is so structured that policy can systematically influence output and employment in the short run, and another thing altogether to say that policy makers have enough knowledge to use that ability in a way that will be beneficial.

(Laidler, 1981, p. 18)

In this chapter, we will set out a formal statement of the contribution which supply-side thinking makes to the way in which we perceive the world and how this applies to regional economies. The next section, therefore, discusses supply-side thinking at the aggregate, macro-economic level. This theoretical discussion is supplemented by the following section, in which the practical application of these ideas will be considered, mainly by reference to the policies pursued in Britain since the late 1970s. In the final section, we will revert to a theoretical discussion, this time of the supply-side approach to regional problems.

Before we then attempt to apply this theoretical approach to real-world regional problems, it will be helpful to review what we know about the locational needs of firms, since this will give some insight into the mechanisms of regional supply responses. For this purpose, Chapter 8 discusses the experience with modern high technology industries, the diffusion of innovations among existing firms, and the process of new firm formation. This experience provides some clues for the practical application of supply-side policies in a regional context, a matter which is taken up in Chapter 9.

## National supply-side

*Theory*

Figure 7.1 takes up the elementary portrayal of the relationship between aggregate supply and aggregate demand which we have already encountered as Figure 3.1. Both diagrams deliberately break the convention that supply and demand schedules are drawn concave upwards. This convention derives from the probability that 'real' schedules are in fact of this character. However, if a 'realistic' shape is adopted, one may be tempted to suppose that it is a faithful representation of reality. By selecting an implausible straight line, we are forced at all points to ask the question: How would the analysis differ if the schedules were of a different shape? The full implications of this point will become apparent as we proceed. Meantime, by way of an example, the Keynesian assumption that supply will respond to demand in an unproblematic way up to the point of full employment implies that a section of the aggregate supply (AS) curve is *horizontal*, an assumption which clearly is at variance with the portrayal in Figure 7.1 and which may, or not, be as plausible.

Suppose that the level of output represented by Q is not sufficient to provide employment for all those seeking work, with the consequence that there is significant unemployment, which it is desired to eliminate. The Keynesian solution is to arrange for an increase in demand, which is equivalent to moving the AD curve to the right ($AD_1$) in Figure 7.1. $AD_1$ intersects AS at $X_1$, giving a total output in money terms of $Q_1$. Whether this nominal output would represent an increase in real output, and hence of employment, depends on the relationship between the change in nominal output and the change in prices – in Figure 7.1 the intersection of $AD_1$ and AS gives a price of $P_1$, which is higher than P. If in fact the supply curve (AS) is horizontal in the relevant area of the diagram, shifting the demand curve to $AD_1$ will cause no increase in prices. On this assumption, $AD_1$ intersects AS (horizontal) at $X_2$, giving a value of output $Q_2$; with no change in prices, the increase in nominal output is exactly matched by the increase in real output. This is the Keynesian assumption.

A symmetrical analysis starts from the assumption that the demand curve (AD) is given. If the supply curve can be moved from AS to $AS_1$, the intersection of $AS_1$ and AD at $X_3$ will again yield a nominal output of $Q_1$, but this time at a lower price ($P_2$). How much of the nominal increase in output represents real output and hence jobs again depends on the shapes of the AS and AD curves. Supply-side advocates stress the possibilities for moving the AS curve in order to increase output and employment, whereas Keynesians put the emphasis on the level of demand.

Logically, control over total output, and hence employment, may be attempted either by the manipulation of demand or of supply, or preferably by both means. The fact that in the short run it is easier to influence aggregate demand than aggregate supply should not blind us to the longer term possibilities of influencing the latter. Furthermore, in so far as either approach has an impact on prices, the direction of price change is likely to be upward if demand is increased and downward if the supply capacity of the economy is improved.

Let us now concentrate on the 'pure' supply-side approach to increasing output and employment. The principle is to shift the AS curve downward to the right. If the resulting changes in nominal output and prices can be calculated, then one can estimate the change in real output that would be occasioned by a given shift in the AS curve. How that change will translate into additional jobs depends on the relationship between the output and employment. It is difficult to estimate this relationship because a shift in the AS curve implies some change in the organization of production, the effect of which will be to economize on labour. However, we can be quite certain that the Keynesian assumption, that employment is proportional to output, represents a limiting situation which is inconsistent with a shift of the AS curve downward and to the right.

Some light is thrown on this problem by the results of empirical enquiry. In the short run, the Organisation for Economic Co-operation and Development countries collectively have experienced employment increases of 0.8 per cent for every 1 per cent increase in manufacturing output (OECD, 1989). In contrast, the long-run studies mentioned in Chapter 2 suggest that increases in labour input account for only about one-quarter of the expansion of output, the remaining three-quarters being contributed by increments to the stock of capital and by productivity improvements. Consequently, we may suggest that the employment elasticity of changes in real output lies somewhere between 0.25 and 0.8, depending on the time period being considered.

Implicit in the discussion thus far is the assumption that we are dealing with a closed economy, in which supply and demand are kept in balance solely through the price mechanism. In a closed economy, this must be so, since income must equal expenditure, unless there is a change in savings. If an economy is in fact open – as all modern industrial economies are – then the adjustment process becomes much more complex than just price and quantity. The non-price characteristics of production have an important role to play. International trade studies, especially in the context of the problems posed by import penetration and poor export performance, show that price accounts for only about one-half of a nation's competitiveness. The other half is attributable to the combined effects of quality, design, delivery dates, the reliability and continuity of supply and after-sales service for those products where this is relevant (Chisholm, 1985a, p. 308). Consequently, the scope for shifting the supply curve to the right depends not just on the reduction of unit costs but also on the ability to cater for customer needs in the other dimensions of competitiveness as well. Very often, improvement in non-price competitiveness can be obtained with fairly modest capital investment, being dependent on such things as the efficiency of the production flow system, the quality control of products and the willingness of the workforce to avoid disruption through strikes (which is in part conditional on the calibre of management).

To shift the supply curve downward and to the right is to innovate – through capital investment and through changes in work organization. In general terms, these are process innovations, introduced in the context of existing production of familiar goods. In addition, and more radically, the supply capability of an economy is affected by the scale of product innovation, the introduction of new goods and services. Consequently, there are two main strands to supply-side thinking. The first strand focuses on market rigidities which impede the continuing (but not necessarily continuous) reallocation of resources and consequently hinder innovation in work practices. The second emphasizes the entrepreneurial aspect of both process and product innovation.

At the heart of supply-side thinking is the view that the technical coefficients in production not only must but do change, if not in a continuous manner then at least frequently and to a degree that is significant. If these changes are to occur at a rate commensurate with circumstances, all economic actors must be faced with appropriate incentives, and these incentives must be mutually consistent. Therefore, supply-siders maintain, close attention must be paid to the pattern of incentives facing firms and individuals, and they complain that the neglect of such considerations was a major failing of the Keynesian analysis.

The simple neo-classical view of incentives is that 'the market' should be allowed to operate freely. Producers and consumers would receive the 'appropriate' signals, in the form of prices, and would respond to those signals. Under these circumstances, the requisite supply would be forthcoming. There are several reasons why it is impossible to rely exclusively on this approach. The monopolistic competition critique of the neo-classical view of the world compellingly shows that if some semblance of a truly competitive situation is to be achieved and maintained, some degree of regulation will be a continuing necessity. Furthermore, if some approximation to fully competitive markets can be achieved, the interpersonal distribution of income and wealth may be judged to be inequitable, creating a situation in which further regulation or the management of transfer payments, or both, will be necessary. The third reason, elaborated by Clark *et al.* (1986), is that the adjustment process is never perfect, and the fourth is the fact that cultural differences between nations result in differing responses to economic stimuli, with the implication that the concept of a 'market economy' is in part conditioned by the part of the globe being considered (Hedlund (ed.), 1987).

Another limitation of the 'market economy' approach arises from the existence of externalities. These lie at the heart of the present concern about global environmental change. For example, no country is willing unilaterally to limit the emission of carbon dioxide, since such self-denial will avail it little if overall the atmospheric concentration of carbon dioxide continues to rise on account of action (and inaction) elsewhere. There is a divergence of national and international interests which can only be reconciled by regulation or by the creation of an artificial 'market' – by the issue of pollution permits which can be bought and sold – since there is no natural market which will achieve the limitation of emissions.

Important externality effects also exist in the regional context, reinforcing the point that governments should not rely solely on creating some semblance of a free market economy. As will be argued later in this chapter (p. 135), the existence of regional externalities (positive and negative) implies the need for some form of intervention, by central government, by designated agencies or by local government – or by some combination of all three.

For the present, though, two points deserve emphasis. First, if the supply side of the economy is to respond with reasonable speed and in an appropriate manner, it is essential that there are suitable incentives for all economic actors. On the other hand, it is not sufficient to rely solely on 'the market', since, at the very least, monopolistic tendencies must be curbed and externality effects must be taken into account. Consequently, a supply-side approach to the economy is not equivalent to the neo-classical *laissez-faire* view, despite the emphasis on getting market signals 'right'.

If the central idea of supply-side thinking is to shift the AS curve downward and to the right, we now need to consider how this may be done. In the next section, therefore, we review some of the steps which have been taken at the national level in recent years.

## *Practice*

By the mid-1970s, there was a widespread view in the industrial nations that an improvement in economic performance and hence in standards of living could not be achieved unless important changes occurred in supply-side conditions. In particular, it was widely accepted that important rigidities in market conditions should be ameliorated, or preferably removed altogether, and that the pattern of incentives was often inadequate. However, the variety of national circumstances (Hedlund, 1987) meant that there was no single blueprint for action. Among the major industrial nations, it is probable that in the 1970s Britain suffered from supply rigidities which were as great as, if not greater than, those found elsewhere. Certainly it is in this country that, from 1979 onwards, the most radical steps have been taken with the aim of creating a more 'flexible' economy. Therefore, this country provides a range of experience which illustrates the nature of supply-side policies, designed to shift the aggregate supply curve downward and to the right. However, the account which follows should not blind anyone to the fact, first, that political rhetoric is not always matched by deeds, and, second, that the deeds themselves have not addressed all of the obvious supply-side problems. The British economy has achieved a marked improvement in manufacturing productivity during the 1980s, generally across all industrial sectors, and the rate of improvement has exceeded that of many competitors. Yet unemployment, though falling, was still over 1.6 million toward the end of 1989, inflationary pressures were serious, the balance of payments was heading for a record deficit of £20 billion and there was considerable pressure on the value of sterling. To combat the inflationary tendency and the pressure on sterling, interest rates were raised to 15 per cent. Some critics (e.g. Green (ed.), 1989) conclude that the supply side of the economy has not been significantly restructured.

The most prominent measure designed to improve work incentives has been the reduction in the marginal rate of income tax from 83 per cent to 40 per cent.

Especially for those on high incomes, a larger share of marginal earnings is now retained by the individual. This should raise the marginal utility of work relative to leisure, which should (according to supply-side thinking) increase the incentive to work. Whether an individual can in fact respond in this way depends on the opportunities for working extra hours at his main job and/or finding a second employment. Nevertheless, as noted on p. 55, there is some evidence that the intended effect has been experienced in practice. The standard rate of tax has also been reduced, from 33 per cent to 25 per cent, and the income band chargeable at this lower rate has been widened.

However, these changes do little for those whose incomes fall below the tax threshold, whether that income is derived from earnings or welfare payments. Many people on low incomes are caught in what is known as the 'poverty trap'. If a person moves out of unemployment into employment, or from a job with low pay to one with better remuneration, three things happen. First, some or all of the welfare payments are withdrawn; second, National Insurance contributions may become payable; and third, tax may become due at 25 per cent on marginal earnings. The way the rules are written and applied, many individuals find that for each extra pound they earn they lose, in benefits withdrawn and in taxes plus National Insurance levied, at least as much and possibly more. The marginal 'tax' may approach or even exceed 100 per cent (Kronsjö, 1987). Under these circumstnces, there is little incentive to effort and self-help.

Reduction of the standard rate of tax to 25 per cent provides some amelioration of the poverty trap. In addition, the rules in respect of National Insurance contributions have been modified, but it was only in October 1989 that the biggest anomaly in National Insurance was removed. Until that date, no contribution was payable on weekly earnings of up to £43, but if earnings rose above that threshold the full contribution had to be paid on all income including the first £43. Even with the October 1989 modifications, the other major cause of the poverty trap remains virtually untouched, namely, the effect of means-tested welfare payments. If earned income rises, benefits will be withdrawn and the recipient can find that virtually all his extra earnings disappear as he ceases to be eligible for welfare payments. The government has shown no sign of adopting the radical solution of incorporating welfare payments into the tax system, treating such payments as negative taxes (i.e., credits) and ensuring that at all income levels the net effect of marginal changes is to give individuals an incentive to earn more. Instead the government seems to take the view that statutory minimum wages and reasonably generous unemployment and other benefits put a floor on wage rates. Consequently, it is held, wages do not adjust downwards in real terms, with the result that potential workers are priced into unemployment (Minford *et al.*, 1983). This line of reasoning leads to the policy conclusion that the wage floor itself should be pushed downward, by reducing the real value of welfare benefits, by curtailing the scope of wages councils to fix minimum wages and by expanding short-term provision of training-cum-employment schemes.

Other countries have wrestled with the problem of incentives for people on low incomes, especially the unemployed. Many American states operate work-fare schemes. The basic principle of these schemes is that welfare entitlement for anyone who is out of work depends on the individual engaging in active, full-time search for work; or on following a course or skill training; or accepting such work as is assigned to them. An individual who is not willing to accept one of these conditions will not qualify under the workfare programmes which exist in twenty-three states (*Times*, 7 April 1986). Another approach, adopted most notably in Sweden, is to provide very intensive help to individuals who are looking for work. Each member of staff in the employment offices has only 22 unemployed people to assist, compared with 82 in West Germany, 193 in Britain and 270 in France (*Economist*, 26 November 1988, p. 24).

A recent change of major significance in the British context concerns the treatment of income from savings. During the early post-war period, it was widely held that 'unearned' income was a source of social inequality and was immoral. 'Unearned' income was consequently taxed at penal rates – the top tax rate was 98 per cent. Some modest exceptions were permitted, most notably National Savings. With much higher taxes on 'unearned' than on 'earned' income, individuals were encouraged to use their income for immediate consumption; the deferment of gratification, through savings, was discouraged. This bias has now been rectified, so that income from savings is treated for tax purposes in the same way as other recurrent income. At the same time, capital gains tax has been reformed, removing a major anomaly. The tax used to be levied on the whole nominal increase in the value of an asset at the time of its disposal. Since some part of that nominal increase would be due to general inflation, tax was being raised on a larger sum than, on any reasonable concept of equity, it should have been. In addition, inheritance tax rules have been modified so that small family businesses can be passed to the next generation without the imposition of a crippling tax burden. All of these changes are designed to give individuals a greater stake in the long-term future of the economy than was the case under the previous arrangements, which had been designed with egalitarian principles uppermost.

The post-war period saw a considerable increase in the power and privileges of trade unions, not only in Britain but also in North America, Australia and elsewhere. Part of the reason lay in Keynesian economic management, under which there were considerable advantages in being able to conclude national wage agreements for all workers in a particular industry. For this to be possible, strong unions were necessary, able to bargain on behalf of their members and able to deliver to employers their members' agreement to the deals that were struck. However, the advantages conferred by such arrangements were less obvious when the consensus about Keynesian management collapsed and in the light of the growing sense that unions, concerned with the welfare of their existing members, stood in the way of changes which were of more general benefit. When Mrs Thatcher came to office in 1979 and Mr Reagan in 1980, it was widely believed that union power had become excessive and that steps must be taken to find a better balance between the rights of organized labour, the rights of individuals and of firms, and the national interest.

Since then, considerable steps have been taken to reduce union power, though not without a certain amount of conflict – for example, the dramatic nationwide sacking of air traffic controllers by Mr Reagan and the bitter 1984 strike by the National Union of Mineworkers in Britain. In the British case, successive Acts have eliminated the right to engage in secondary picketing, have imposed the necessity for ballots before official strike action and have put union assets at risk if a union breaks the law. The idea that unions should be subject to the due processes of law was initially anathema to many unionists but seems now to be generally accepted, even welcomed by many members and unions, though differences of opinion will always remain concerning the precise balance of rights and powers. In addition, the government has sought to encourage movement away from national wage agreements to local, even plant-based deals on wage rates and the conditions of work, though it has proved harder to achieve this in the public sector than in the private sector.

Whatever damage some union practices may have inflicted in the past, it would be wrong to conclude that trade unionism necessarily is in conflict with the need to innovate to obtain productivity increases. Thus, taking the wider context as given, over the period from 1975 to 1986, British companies which had a highly unionized workforce achieved productivity increases matching the increases in less unionized companies (*Independent*, 29 August 1989). Trade unions are quite powerful in both West Germany and in Sweden, two countries whose economic performance in the post-war period has been enviable. It is probable, therefore, that it is less a question of the degree of unionization and more a matter of the way in which unions operate that determines their impact on the ability of companies to adapt and innovate.

Reasonable success has attended efforts to encourage wider ownership of shares and the adoption of profit related pay deals. Depending on the basis of measurement, whereas in 1979 only 5 per cent to 7 per cent of the British population owned shares, by 1987 the proportion had risen to between 18 per cent and 19.5 per cent (*Economic Progress Report*, No. 189, 1987, p. 2). Profit-related pay schemes have been advocated by Weitzman (1984) and were given a considerable boost in the 1987 Budget: the main effect, however, seems to be to raise productivity rather than, as Weitzman believed, to reduce unemployment (*Economist*, 25 April 1987, p. 38). More generally, it does seem likely that increased share ownership and the spread of profit-related pay will help to create the circumstances in which individuals have a clearer stake in the success of their employers, while firms will be faced with wage bills that are less sticky in a downward direction than has hitherto been the case, or has been assumed in the Keynesian frame of reference.

These moves in respect of wages and profit-related pay are consistent with recent British studies which indicate that the elasticity of labour demand with respect to real wages lies in the range of $-0.5$ to $-0.1$. With total employment at about 21.5 million, these elasticities imply that a reduction in real wages of 1 per cent would result in higher employment of between 110,000 and 220,000 (Treasury Officials, 1985, p. 24). The precision of these estimates, which are derived from the extant published work, may be open to question; the nature of the relationship, however, lends support to the direction which policy has followed in recent years.

Pay cannot sensibly be divorced from the nature of the work to be performed. A marked feature of the last decade has been the move by many employers to reduce the number of separate job descriptions and thereby to make their workforce more flexible. Very frequently, moves in this direction are linked with pay negotiations. For example, in 1985 Ford UK offered a pay settlement to its 37,000 manual workers that was conditional on their accepting changes in job demarcations which would reduce 500 specifications to 58, with the aim of raising overall productivity toward the standards prevailing in West Germany and Japan (*Times*, 27 November 1985). Unfortunately, the Dagenham site, at least, is very congested and ill-planned, making efficient working difficult to achieve; although productivity in this plant has improved, the level remains lower than that achieved across the Channel, and Ford may decide to close it (*Independent*, 12 October 1989). Deals concerning work practices and manning levels are difficult to achieve even if circumstances are propitious. If more than one union is involved, and if one of these believes that its members' interests would be jeopardized, serious conflict can erupt. This happened in the newspaper printing industry, as the National Graphical Association (NGA) – the typesetters – sought to prevent the direct keying of text by reporters, claiming the right for NGA members to do this work, notwithstanding that this would involve unnecessary labour given the availability of modern word processors. The issue was not really resolved until 1985, in favour of direct input by reporters, by-passing the NGA. Difficulties of this kind can be avoided under single-union deals, but these can usually be obtained only when a brand new plant is being built from scratch.

If labour markets are to be responsive, workers must be able to move geographically. In the British context, there are two major impediments to such mobility which are associated with the housing market. Just under one-third of all dwellings are owned by local authorities and rented to their occupants. Compared with owner-occupiers, who account for nearly two-thirds of dwellings, council tenants have a low propensity to move between standard regions. The primary reason for low mobility is the difficulty experienced in moving from a local authority dwelling to one owned by another authority without a long waiting period (Hughes and McCormick, 1981). The government has sought to reduce this impediment to mobility by introducing the 'right to buy'. Council tenants can buy the freehold at very advantageous terms. About 15 per cent of the council stock has been sold in this way in the last decade.

Although owner-occupiers are considerably more mobile interregionally than are council tenants, both groups are much less mobile than individuals in privately rented accommodation. This sector accounts for only 8 per cent of all dwellings. Although the government has recognized the desirability of expanding the supply of private rented accommodation, relatively little has been done. Provisions are in place for the 'right to rent', namely, the conversion of council tenancies into private rental associations, but little has been done to reduce the high level of legal protection which private tenants often enjoy, and nothing to redress the imbalance caused by the combination of public subsidy to council tenants and mortgage tax relief for owner-occupiers. Private rents are comparatively expensive, making this form of tenure unattractive or unavailable for many who would otherwise value the freedom conferred by private rental arrangements.

If the pattern of personal incentives and the operation of the labour market are important, so also is the environment in which firms operate – does it foster or stifle enterprise, does it encourage efficiency and innovation? The basic aim of government policy during the last decade has been to increase the element of competition in all sectors of the economy, in the belief that this will lead to greater efficiency. Several strands can be readily identified. First, in the belief that state-owned industries are less efficient than private companies, major industries have been transferred from public ownership in a process known as privatization, a process which includes flotation on the stock market and inducements to members of the public to acquire shares. Gas, the telephone system, British Airways and British Petroleum (BP), for example, have all been privatized, and similar treatment is in hand at the time of writing for both electricity and water. Whether genuine competition is actually engendered by privatization is a moot point. British Airways competed with other airlines before privatization, as did British Petroleum with rival petroleum companies. British Gas has been privatized as a monopoly supplier of gas, but in competition with other fuels (mainly electricity and petroleum) for markets. The privatized water industry will consist of regional companies, monopolists within their areas, selling a commodity for which there is no substitute: competition is not relevant in this case. Perhaps the main effect of privatization has been the removal of the Exchequer safety net, the sense that there is no refuge for the enterprise which fails to maintain and improve efficiency.

Other steps which have been taken include the insistence that local authorities and other public bodies put work out to competitive tender and the idea that some schools and hospitals may opt out of the public sector. In addition, the British government, following the lead of the American and Canadian governments, has been pressing for the deregulation of European air services. Finally, there have been attempts to reduce the general bureaucratic burden on firms and the development delays caused by planning procedures. The Department of Trade and Industry published a White Paper in 1985 on the former issue, as a declaration of intent. More tangible, perhaps, has been the series of circulars to local authorities concerning the interpretation of planning laws and procedures, designed to make it easier for development to proceed, but this trend has raised acute anxieties concerning the quality of the environment.

The preceding paragraphs have summarized the major recent policy changes in one country, the United Kingdom, designed to improve the supply side of the economy. The list of measures mentioned is incomplete but sufficient to indicate the practical possibilities for shifting the national supply schedule to the right and, it is hoped, facilitating the emergence of new firms and new products. Despite the changes which have been implemented since 1979, the British economy probably remains a good deal less flexible than is the case with several major competitors. With 42 per cent of workers being union members, centralized pay bargaining remains much more important than in the United States, where only 18 per cent of workers belong to a union (*Economist*, 19 August 1989, p. 15). There is greater real-wage rigidity in Britain than in other advanced economies, which may be a partial consequence of high levels of unionization, or, more probably, of basic attitudes to wage bargaining. Wage

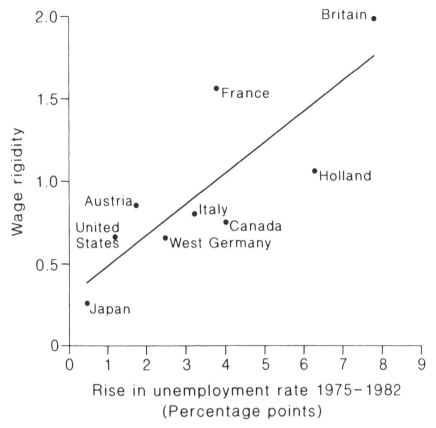

**Figure 7.2**   Wage rigidity and unemployment, 1975–82
Wage rigidity is a measure of the extent to which wages respond to prices
rather than unemployment
*Source: Economist,* 15 June 1985, p. 69

rigidity in this context means that wages respond to prices rather than to the
level of unemployment, as workers seek to maintain their standard of living in
the face of inflationary price changes (Figure 7.2). Another element of rigidity
which remains very marked is the low mobility of the working population, at
least in comparison with the United States. There, manual workers are eighteen
times more likely to move from one state to another, either to find or to keep a
job, than are their British counterparts to move from one standard region to
another (*Economist,* 19 August 1989, p. 15).

On one interpretation, therefore, a good deal remains to be done to create in
Britain the flexibility of national supply responses which is required. The
experience of 1989 – a record external deficit plus serious inflationary pressure,
combined with unemployment well above any plausible definition of full
employment – suggests that the achievements to date have indeed been insufficient. Perhaps the most obvious failure has been in respect of education and

skills, to equip workers for modern jobs and to facilitate retraining to cope with changing employment needs. As an editorial in the *Independent* (4 May 1989) pointedly observed: 'Above all, Britain's economic future will depend upon the skills and abilities of our people and it is these which Mrs Thatcher has shamefully neglected.' The transport infrastructure has also been seriously neglected in comparison with the other member countries of the European Community (Confederation of British Industry, 1989). In both cases, of course, the need is for additional public expenditure, which the government has been reluctant to find, being pledged to reducing the share of GNP which passes through the Exchequer.

The British government has meantime stressed the significance of the pending creation of a 'single European market' in 1992. The intention is to remove physical, technical and fiscal barriers to trade, so that the national markets of the states comprising the European Community will be more genuinely open than is currently the case. More than 300 separate measures are envisaged, ranging from the abolition of exchange controls, through the removal of restrictions on international road haulage to the mutual recognition of higher education diplomas. These changes are projected as a major development of supply-side policies, as indeed they prospectively will be. The scale and the nature of the impact is, however, still a matter for speculation. What is not a matter of speculation is the acceptance by the leader of the Labour Party, Mr Kinnock, of the urgent necessity to foster the productive potential of the country. In his keynote speech to the Labour Party conference, in October 1989, he described education and training as 'the commanding heights of every modern economy', and his speech was described in an editorial as a programme for the Labour Party to 'concentrate on supply-side deficiencies of the British economy' (*Independent*, 4 October 1989).

Over the last decade, there is no doubt that some British regions have been hit particularly hard, while others have been relatively prosperous (Martin, 1988a, 1988b, 1989). The question now arises whether such shifts in relative regional fortunes, which have been experienced in most countries, are something which happen to regions in an exogenous manner? Or, to what extent, if any, do the workers and inhabitants of a region have any control over the course of events? To begin to answer that question, we need to consider the light that supply-side theory can throw on regional growth processes. In particular, can regional problems be formulated in supply-side terms similar to those which are appropriate at the national level?

## Regional change

*Theory*

Figure 7.1 shows how, at the national level, output (and hence employment) could be increased if the aggregate supply curve (AS) could be shifted downward and to the right. This idea can be applied quite directly to analysis at the regional level. Imagine that a homogeneous commodity is produced in two regions which are competing in the same market. Let us also suppose either that

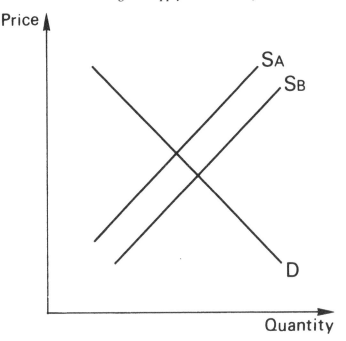

**Figure 7.3** Two regions in competition

differential transport costs from supplier to market can be ignored, or that the supply schedules include distribution costs. If, furthermore, we suppose that there is only one firm in each of the two regions making this homogeneous good, then 'region' and 'firm' become interchangeable for certain purposes. To keep the discussion clear, however, we will approach the problem of interregional competition from the perspective of the two imaginary firms, respectively A and B.

In Figure 7.3, both firms face a common demand schedule, D. However, firm A has higher production costs than firm B, as represented by their respective supply schedules ($S_A$ and $S_B$). Under the circumstances portrayed, firm B can supply the whole market at a price below that required by its competitor; firm A would be in danger of going out of business.

The two-firm, two-region case is an extreme situation. More usually, there are several firms in competition, commonly located in numerous regions and with some degree of spatial monopoly conferred by transfer costs. Nevertheless, high-cost producers of a homogeneous commodity are at risk from the competition offered by producers whose costs are lower, and the two-firm case can be used as the basis for discussion. To avoid the fate of going out of business, our firm A will need to consider the strategies which are open to it – which of them to pursue, or in what combination?

(1)  To seek protection in one form or another (e.g., tariffs, subsidies)
(2)  To seek a collusive arrangement with the more efficient producer to share the market

(3)   To seek to lower production costs to match the supply schedule of B, the more efficient producer
(4)   To seek to differentiate its product and carve out a new market
(5)   To seek to move into the manufacture of other commodities.

The first of these options may provide the breathing space which is necessary to follow one or more of the other options, or it may be deemed necessary in the longer run to counter 'unfair' competition which may arise through the payment of export subsidies to competitors, or other assistance. We will regard protection as a permissible short-term measure but not one to pursue on a long-term basis. Collusion, the second possibility, is apt to be against the general public interest and, therefore, will also not be discussed any further. The three remaining options are all constructive responses to the competitive situation which has been outlined and deserve further scrutiny.

LOWER PRODUCTION COSTS
The problem of lowering production costs and hence bringing the $S_A$ schedule down to or below the $S_B$ schedule can be approached in one of two ways: relocation of the firm, to a place where operating costs are lower – with cheaper materials inputs, lower wages, greater ease in organizing the workforce, etc. – or, alternatively, *in situ* changes. It is the latter upon which we will concentrate.

At a time of crisis, as in a sharp recession, it may be possible to renegotiate supply contracts and obtain the agreement of workers to take a cut in wages. In adition, a firm may accept a sharp reduction in its profits, even a loss, in order to hold prices down. These are short-term measures which do not address the underlying structural problem posed in Figure 7.3. The key issue, if the supply schedule is to be shifted downward, is how to increase the efficiency of the production process. This amounts to the need for process innovations, which run the whole range from capital investment with state-of-the-art technology, through the reorganization of management systems, to changes in shop-floor practice. Since wages historically make up a large share of total production costs, it is usual to seek cost reductions by reducing the workforce needed for a given level of output. To achieve this, it is common for capital investment to be necessary. However, it may be possible to reduce overall costs in the absence of new investment, by, for example, operating two or three shifts instead of just one. Even if capital investment is a necessary condition for achieving lower costs, it is seldom or ever a sufficient condition; changes in manning levels and work practices will also be necessary. This lesson was learned the hard way by General Motors. Their truck factory at Hamtramck, Michigan, has lower productivity and poorer quality control than their plant at Fremont, California; the former plant has had the benefit of heavy investment in robots, the latter is run on labour-intensive lines in a joint venture with Toyota (*Economist*, 21 May 1988).

To shift the supply curve downward is to engage in process innovations, of which numerous examples can be cited. In his study of the metalworking industry in the United States, Hicks (1986) unequivocally found that the primary reason why firms installed numerically controlled machines (computerized or otherwise) was to increase productivity and quality, in order to improve competitiveness. Another survey of American firms, this time in the machinery manufacturing sector, found that the adoption rates for six out of eight innovations studied were significantly higher for the older plants than for the newer ones, as reinvestment took place to maintain competitive positions (Rees *et al.*, 1986). In the early 1980s, steel producers in both the United Kingdom and the United States faced ever mounting financial losses and continuing erosion of domestic and overseas markets. In both cases massive reorganization and reinvestment has turned the situation around. British Steel, the world's fourth largest producer, had by 1988 become the most profitable of the large manufacturers, albeit with the help of considerable public expenditure during the transition period. At the same time, the American industry re-established itself as the cheapest supplier in its domestic market, as shown in Table 7.1. In both cases, major investment was required, and a savage cut in the workforce:

|  | Steel workers | |
|---|---|---|
|  | USA | UK |
| 1980 | 400000 | 125000 |
| 1987 | 155000 | – |
| 1988 | – | 50000 |

In addition, there were considerable organizational changes, including worker buyouts of several plants in the United States, with the backing of United Steelworkers of America (USWA), and associated wage cuts (*Economist*, 16 April 1988, p. 49; 17 December 1988, p. 75). Less spectacular adaptations have been reported for other lines of business. Examples include industries generally regarded as being high technology, in which output is growing, such as the computer industry and Philips Electronics, as well as the more mundane food

**Table 7.1** Selected countries, cost of steel per tonne, landed in the United States, November 1988

| Producer | Materials | Labour | $<br>Capital[1] | Shipping[2] | Total |
|---|---|---|---|---|---|
| United States | 291 | 160 | 33 | 0 | 484 |
| United Kingdom | 295 | 104 | 21 | 71 | 491 |
| South Korea | 272 | 51 | 95 | 80 | 498 |
| Taiwan | 275 | 72 | 75 | 80 | 502 |
| France | 280 | 125 | 49 | 73 | 527 |
| West Germany | 288 | 127 | 58 | 72 | 545 |
| Japan | 297 | 150 | 105 | 80 | 632 |

[1]Depreciation and interest.
[2]Freight and customs duties.
*Source: Economist*, 17 December 1988, p. 75

and drinks produced and marketed by Cadbury-Schweppes (Kelly, 1987; Peck and Townsend, 1987).

PRODUCT DIFFERENTIATION

An alternative way of cutting production costs is available for industries in which scale economies are significant, namely, energetic marketing to expand sales. If such a campaign is successful, overhead costs will be spread over a larger total output, cutting unit costs. Such a campaign will usually involve persuading customers that the product is different from and better than rival products. This amounts to an attempt to differentiate the product in the mind of the customer, even if objectively there is little or no real difference. Generally, though, a marketing campaign will be based on some degree of real product differentiation, which in turn will require some process and product innovation. Product innovation is widespread, as firms seek to identify market niches which are relatively sheltered (monopolistic), or those niches in which competition is fierce and the struggle is on for customer loyalty, with big potential rewards for the successful firm(s).

Part of the reason for the decline of Britain's shipbuilding industry must lie in the slow response to market needs and the failure to find specialist lines of production which could be exploited – without the need to build supertankers in estuaries too small to accommodate them – in the way that Finland found its niche with icebreakers. The demise of the British motorcycle industry after the Second World War owed much to the failure of the manufacturers to recognize the trend away from big machines of 350 cc and 500 cc capacity to much smaller and lighter machines.

NEW PRODUCT

Product innovation is a continuing process, in which many firms engage as a matter of course as part of their normal business. New pharmaceutical and chemical products are constantly being brought to market; the market for recorded music has witnessed the eclipse of traditional records by tapes, which in turn are threatened by compact discs. Radio has been supplemented by television. The list of examples would be endless. For a firm such as A in Figure 7.3, if cost cutting and/or product differentiation prove inadequate, then the third option is to enter an entirely new market. One of the more dramatic examples of this being attempted was the effort some British shipyards put into obtaining orders to build rigs for the exploration and development of the North Sea gas and oil fields, an effort which was not in fact notably successful in practice. Much more successful was the transformation of many firms in the Birmingham/Black Country area early in the twentieth century: as the market for traditional foundry and brass products declined, they moved into the business of supplying components for the burgeoning car industry.

## Research and development

If lower production costs, product differentiation and the introduction of new products are the proximate means by which the supply schedule can be shifted, these strategies are often dependent upon product and process innovations which arise from the application of science to production problems. Therefore, underlying all of the discussion thus far is the activity that is usually called R and D, which may be carried out within a company, in separate research establishments funded by either the private sector or the public purse, or in higher education institutions such as universities. In the present context, we need do no more than note the basic role of R and D in the process of technological change: its significance for regional development will be examined in Chapter 8.

## The firm and its environment

If a firm is to shift its supply curve downward, if it is to differentiate its product or if it is to shift to a new product(s), then the essence of the matter is the need to innovate. At one level, decisions about process and product innovation are a matter for the firm itself, in which case the region within which the firm is located is a passive actor with no power to intervene to affect the course of events. In practice, the freedom which a firm has to make the decisions which it deems to be necessary is constrained in a number of ways. This is portrayed in Figure 7.4. If we imagine that a firm has an action space, some part of that space overlaps with the action space of other agents, as represented by the shaded area. The external agents may be national (e.g., national laws and regulatory agencies) or local (for example, local government and regional development agencies).

Both monopolistic competition theory and the Keynesian view of the world draw attention to the existence of externalities, which, by assumption, do not exist in neo-classical theory. Because firms exist in a spatial framework, they are necessarily affected in some degree by conditions external to their own

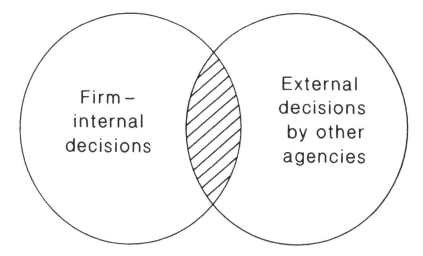

**Figure 7.4**  The firm and its decision environment

operations, just as much as they also have an impact thereon. Local govern-ment jurisdictions provide public goods and services, such as roads and site services, transport, education and training facilities, etc. Many workers are members of trade unions, so that decisions about pay, conditions and work practices must be taken in a context wider than that of the firm alone.

A firm which wishes to innovate may well find that its own decisions must be contingent upon decisions taken by other agencies at the local and/or national levels. For this reason, it becomes a matter of considerable interest to examine the nature and the importance of the area in which, as shown in Figure 7.4, the firm's action space overlaps with the action space of other agencies, and especially with those which operate locally. The larger the area of overlap, and the more important that it is for the firm, the greater the impact which external agencies may have upon the decisions taken by firms – an impact which may help or hinder the intended innovation (adjustment) process.


## Conclusion

Earlier chapters have shown that both the neo-classical and the Keynesian schools of thought pay little real regard to supply-side issues. Both tend to assume that supply responses will be adequate and that they are therefore unproblematic. In the present chapter, we have explored this missing ingre-dient, in terms of basic theory and practice at the national level, and also theory at the regional scale of consideration. What the balance should be between reliance on neo-classical adjustment processes, intervention to manipulate demand and concern with supply-side issues is a matter over which debate will, no doubt, continue. What cannot reasonably be disputed is that the supply side does matter – especially in the context of the long run – and that supply adjustments cannot be assumed to occur in an unproblematic manner.

In taking up supply-side matters, we have moved from macro-economic questions to the micro-economic behaviour of firms and of agencies which operate at the local level, as well as introducing in a more explicit way the role of individuals. In particular, a firm that wishes to engage in a particular strategy to cope with the competition it faces is not a free agent; its decision space intersects with the decision space of central government, of local agencies and of its employees. So, even though entrepreneurship must be given a prominent place (Schumpeter, 1943), so also must consideration be given to the require-ments of entrepreneurs and of firms. To the extent that these requirements are understood, it may be possible to identify the ways in which the environment facing existing and potential firms can be modified in a manner deemed beneficial to particular regions. Consequently, in the next chapter we will examine the location needs of modern firms and then, in Chapter 9, we will discuss the scope which exists for local and regional agencies to intervene beneficially.

CHAPTER 8

# The location needs of modern firms

> Wouldst thou – so the helmsman answered, –
> Learn the secret of the sea?
> Only those who brave its dangers
> Comprehend its mystery!
>
> H. W. Longfellow, *The Secret of the Sea*

In the previous chapter, we have noted that firms are not entirely free agents when they make their investment and other decisions. Therefore, the decisions taken by local government and other agencies, affecting a locality or region, will have some impact on the choices which firms make. If that impact is to be consciously beneficial, it is essential to know what it is that firms need of the environment in which they operate. Such knowledge, however, will always be imperfect, dependent on an element of inference. Survey data of firms may elicit the opinions of key people regarding the reasons which led to a firm locating and expanding in one place in preference to another. However valid such opinions may be for the surveyed firms, and however well supported these opinions may be by other evidence, there is always the uncertainty that other firms may be affected by other considerations. In contrast, statistical exercises using census or similar data are faced with the problem that high levels of spatial association imply causal connections which are not directly observed and which may be erroneous.

Nevertheless, there is now a large body of evidence which has been published, using a variety of research frameworks, and yielding a fairly consistent pattern of conclusions. Therefore, the main purpose of this chapter is to review this evidence, to see what we may learn regarding the needs which firms have. To the extent that these needs can be met only by agencies external to themselves, there will be scope for these agencies to have a material impact on existing local firms, potential inward migrants and potential new start-ups.

This discussion will concentrate on manufacturing industry, largely because there is far more material on which to draw than is the case with service activities. Within the manufacturing sector, attention will focus on three matters in particular: the location needs of 'high technology' industries; the diffusion of process and product innovations among existing firms; and the setting up and success, or otherwise, of new firms, most of which are small. These three topics encompass most of the supply options available to a region which is faced with the decline of major sources of employment.

## High technology industries

High technology industries have attracted a very large amount of attention in recent years, as witness the following literature (Amin and Goddard, 1986; Aydalot and Keeble (eds), 1988; Breheny and McQuaid (eds), 1987; Brotchie *et al.*, 1987; Chapman and Humphrys (eds), 1987; Gillespie (ed.), 1983; Hall and Markusen (eds), 1985; Hall *et al.*, 1987; van der Knapp and Wever (eds), 1987; Markusen *et al.*, 1986; Oakey *et al.*, 1988; Rothwell, 1982; Rothwell and Zegveld, 1985; Scott, 1986; Thwaites and Oakey (eds), 1985). This and other material has been very usefully reviewed and summarized by Milne (1989) and C. Thompson (1989). Nevertheless, there is no general agreement regarding the definition of what constitutes high technology industry. The general concept is of industries at the leading edge of technological advances, generating new or much improved products, resulting from the application of R and D. Definition could be based on the number of innovations produced or used, on the R and D expenditure of firms, on product sophistication or on the proportion of employees deemed to be highly qualified and/or 'scientific' in their orientation. Most observers include electronics and computing among high technology industries, as also modern communications, many branches of defence production, aeroplanes and biotechnology. But how much longer should the list be? The nature of this difficulty is clearly indicated by a tabulation of nine different definitions of American high technology industry, giving a total employment in the early 1980s ranging from 2.5 million to 12.6 million, the median value being 5.3 million (de Jong, 1987, p. 38).

The most readily available data are for employment, with the result that an employment criterion is commonly used as the basis for defining higher technology industry. For example, Markusen *et al.* (1986), in their careful study of the United States, start with the three-digit industries of the SIC classification. To qualify as being high technology, the proportion of employment in an industry accounted for by 'scientific' employees must exceed the national average. This category of employment comprises the following: engineers plus engineering technicians, computer scientists, scientists and mathematicians. On this basis, 29 three-digit industries were identified as qualifying, comprising a total of 100 four-digit industries, on which the study is based. These industries include:

   radio, TV transmitting, signal, direction equipment;
   industrial organic chemicals;
   petroleum refining;
   aircraft;
   electronic computing equipment.

Keeble (1989) used a somewhat different basis for his selection, incorporating the intensity of R and D with the criterion of the proportion of the workforce classified as scientists, professional engineers and technicians. His list of nineteen industries includes:

synthetic resins and plastics materials;
pharmaceutical products;
optical precision instruments;
telegraph and telephone apparatus and equipment;
electronic data-processing equipment.

On the basis of these definitions, which are probably fairly generous, some 27 per cent of the United States' industrial employment is in high technology industries, and 22 per cent in the case of Britain. Important though these industries are, their growth and development cannot possibly solve the employment needs of all the regions within countries such as these two. In any case, there is no simple link between high and low technology status and employment growth. Whereas in the United States, high technology industries collectively account for most of the employment growth in manufacturing, in Britain they have experienced a decline, albeit more slowly than in industry generally (see also Hall *et al.*, 1987). Of the one hundred American high technology industries, only six accounted for half the employment increase, and twenty-eight experienced a decline over the period from 1972 to 1981.

Notwithstanding the common elements of product sophistication and high level of skilled manpower employment, high technology industries are manifestly heterogeneous. This fact should be kept in mind during the ensuing discussion, since it is inherently possible, even probable, that the factors which determine location will vary from one industry to another. But should one even expect there to be identifiable location factors which can be analysed? Thompson (1989) argues that traditional locational determinants theory is not relevant to high technology industries. This theory, in the tradition of Weber, Hoover, Lösch, Isard and others, focuses on cost minimization or profit maximization and emphasizes transport costs as a major determining factor. It is assumed that there is an identifiable production function for each commodity and that each region has its own unique endowment of the relevant factors of production, such that optimal location choices can be 'read off'. This approach, relevant for the location choices for steel plants and similar 'smokestack' industries, is held now to be inappropriate. Thompson reviews other approaches – long wave theory, product-profit cycle theory, production organization theory and socio-spatial system theory. On closer inspection, however, these approaches should more properly be considered as showing why modern manufacturing operates with locational determinants which differ from those which were relevant for earlier industries.

We have already seen that the long wave literature and the product cycle/profit cycle analyses do not directly address the location problem. At best, they suggest that the determinants of location change over time and that these changes may be systematic. The theory of production organization, which Thompson attributes to Scott and Storper (1987), is based on the advantages which accrue for firms from proximity to the other important economic actors and as such is a version of the scale economy approach to regional develop-

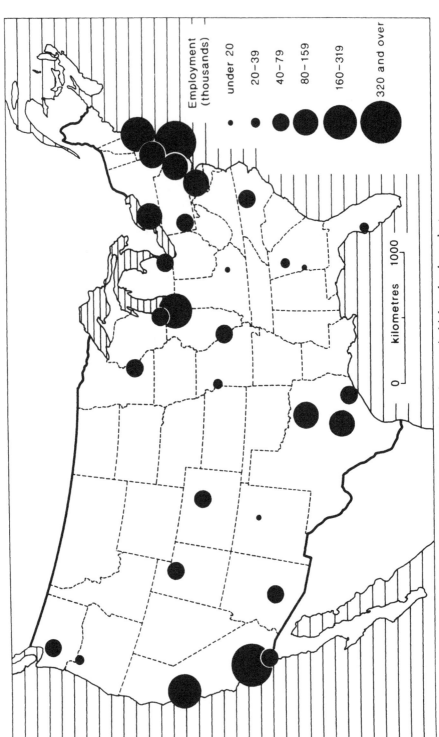

**Figure 8.1** United States: 1984 employment in high technology clusters
*Source*: Miller and Côté, 1987. p. 16

ment, with allowance for the emergence of diseconomies as time passes and as local economies become ossified. Finally, following Castells (1985) socio-spatial system theory, cast in the radical tradition, emphasizes control over information and financial flows, the dualism between the controllers and the controlled and the progressive concentration of headquarters control in a limited number of major cities.

Nobody is suggesting that firms have ceased to be interested in profits, or that location choices are irrelevant to the profitability of firms. All that is being suggested is that the location factors which are significant for competitive success and profitability have been changing, and are markedly different in nature and relative importance compared with those for the leading industries of half a century ago. To examine what these factors may be, we will turn to some of the available evidence based on detailed investigations.

## *Location of high technology industries*

One could be forgiven, reading much of the presently available literature, for supposing that high technology industries are concentrated in a few favoured places, mostly with a limited recent history of industrialization. Silicon Valley near San Francisco, and Orange County hard by Los Angeles, immediately come to mind, as also the M4–M11 crescent in Britain, while the 'Cambridge phenomenon' is a term which is in widespread use (Segal, 1985; Segal, Quince, Wicksteed, 1985). Closer inspection shows, however, that these industries are much more widely distributed than is indicated by recalling the famous examples. This fact is clearly shown by Figure 8.1, which records the distribution of the thirty technology clusters identified by Miller and Côté (1987). The total employment in these clusters in 1984 was 3.3 million, which compares with the total recognized by Markusen *et al.* (1986) of 4.8 million in 1977 and 5.5 million in 1981; the Miller and Côté definition is somewhat more restrictive and is based on localities rather than whole states. The smallest cluster shown in Figure 8.1 has 11,800 jobs and the largest 519,300. Although California has two of the larger clusters of firms, it is striking just how much employment there is in the northeast, and not just in the vicinity of Boston and its well-known Route 128. This point is confirmed by Hall (1987b) and Thompson (1989), and forcibly emphasized by Markusen *et al.*:

The New England example, just quoted, is the most dramatic illustration of the fact that high tech is not a monopoly of the Sunbelt states of the South and West. So are the Chesapeake/Delaware River area and Illinois. Three of the five major state agglomerations of high tech industry, with some 37 per cent of the jobs in them, are thus in the nation's old industrial heartland. At the SMSA level, too, older urban agglomerations – Chicago, Boston-Lowell-Brockton-Lawrence-Haverhill, Philadelphia, Newark, Detroit, New York, Cleveland – actually emerge as among the biggest concentrations of high tech employment. And two of these – Boston and Worcester – were among those ten SMSAs posting the largest job gains in the mid-1970s.

Admittedly, there were only two such; they were dominated in the listings of job growth by the Sunbelt SMSAs like San Jose, Anaheim, Houston and San Diego. And in terms of percentage gains, the records were held by small, fairly isolated urban areas in the interior of the United States, far distant from the major metropolitan areas. Nevertheless, the example of New England does illustrate convincingly that there is no fatalistic rule that the march of high tech is southward and westward away from the old industrial core. Much, still, depends on the ability of an old industrial region – here, literally the nation's oldest – to reconstruct itself. And this ability will turn in large measure on the region's inheritance of physical and social infrastructure and accumulated business services.

(Markusen *et al.*, 1986, p. 173)

Similarly, in Great Britain the southeast (including London) plus East Anglia accounted for 47.5 per cent of high technology industrial employment in 1984, the remaining 52.5 per cent being widely spread through the country (Keeble, 1989; see also Hall *et al.*, 1987). The importance of the southeastern regions was being accentuated, by virtue of a lower rate of job loss there than elsewhere.

Both in the United States and the United Kingdom, the most remarked developments of high technology firms have been in 'new' industrial areas, where the relative growth rate has been favourable. In both countries, however, there is a substantial presence of high technology firms in 'older' industrial regions, a fact which is often given less prominence than is warranted. Britain's 'older' regions have benefited from regional economic policy (pp. 84ff.), but no such explanation can be offered in the case of the United States. The position becomes more complex if account is taken of the geographical distribution of high technology industries in other countries. For example, it seems quite clear that within the Netherlands there is a disproportionate concentration of these industries within the urbanized Randstad region, and within that region the large cities have a more than proportionate share (de Jong, 1987). This distribution seems to provide some contrasts with experience elsewhere.

*Can the location patterns be explained?*

Numerous studies have been published in which an attempt is made to identify the geographical patterns of high technology industry and to explain the patterns thus observed. Individual industries have been subject to scrutiny, such as scientific and industrial instruments (Oakey, 1981) and semiconductors (Scott, 1988b). Other studies have taken the whole range of qualifying industries and sought general explanations applicable to the whole set of industries (e.g., Hall *et al.*, 1987; Markusen *et al.*, 1986). The former group of studies uses a mixture of statistical data and information derived from questionnaires and interviews, while the latter depends very largely on official data in censuses or similar sources. Despite the heterogeneity of high technology industries and the multiplicity of research frameworks (date, geographical coverage, etc.), there is a good deal of consistency in the research findings.

A priori, we would expect the traditional factors of access to fuels, materials and final markets to be unimportant, because there is a large value added in manufacturing. The conventional minimization of transport costs should, therefore, not be important. On the other hand, to the extent that firms depend on components, the specifications of which may change rapidly, proximity between manufacturers at various stages in the production process may be beneficial, to facilitate the personal contact necessary in dovetailing components to their final use. This should confer advantages on areas in which many firms are located, but would not in itself explain why one place got a head start and was then able to build on its success. For that, it seems plausible to suggest the importance of highly qualified manpower – for innovation and development – and skilled labour for production. The dramatic increase in the relative importance of the former has been demonstrated by Green and Owen (1989) for engineering industries in Great Britain. The existing availability of highly qualified and highly skilled people, or the ability to attract them, might then be the crucial element that would allow an initial entrepreneurial decision to trigger continuing growth. If that hypothesis is valid, then we are looking at factors like the residential quality of an area, the amenities that are available and desired by educated and skilled people and the professional environment, including access to research.

Some, but not all, of these expectations are confirmed by the evidence available. The conclusions reached by Christy and Ironside (1987), Hall (1987b), Malecki (1986) and Oakey (1981) suggest both positive and negative findings, as indicated in Table 8.1. The important factors occasion little surprise, being broadly consistent with our a priori expectations, though two points of amplification are essential. To the extent that innovative development is separated from the routine production of standard articles, the labour factor may pull in two directions, with older industrial areas having a clear advantage in the availability of skilled operatives to offset their possible disadvantage with respect to the supply of innovative personnel. The concept of residential amenity is multifaceted, for all the hype about Sunbelt locations. For some people, access to culture may be more important than access to the sun, and for those for whom outdoor pursuits are important, skiing and water-based activities

**Table 8.1** Factors which appear to be important and not important in explaining the distribution of high technology industries

| Important | Not important |
|---|---|
| Skilled labour: innovators | Access to research |
|            operatives | |
| Residential amenities | Industry organization |
| Accessibility (personal): | Cost of transport on goods/ |
|         road system | local linkages |
|         airports | |
| Headquarters function of firms | Unionization |
| Business services | Wage rates |
| Defence expenditure | |

may have a higher priority than high temperatures and an absence of rain. Consequently, the direction in which these important factors are pulling is not necessarily consistent. In any case, as Malecki notes: 'The built environment and cultural amenities are more important than the physical setting of an area' (1986, p. 62). In addition to the quality of the built environment and the cultural amenities, there seems to be general agreement that access to good travel facilities for key personnel is an important locational consideration, as exemplified by the Newbury area west of Heathrow (Macgregor *et al.*, 1986).

The big surprise contained in Table 8.1 is the inclusion of access to research among the unimportant factors, a view which may be disputed by some (for example, Keeble, 1988). Research, in this context, includes both civil and military, the former encompassing universities, independent research organizations and the R and D carried on by firms. The work by Hall and Markusen suggests that proximity to universities is not an important factor, but other scholars (for example, Malecki, 1986) include access to libraries and university personnel among the positive attributes that an area may have. The problem here is that the quality of the resources thus accessible may correlate positively with the more general measures of the quality of the built environment and cultural facilities. Given the manifest importance of R and D for high technology industry, it does seem counter-intuitive to conclude that access to research is an unimportant location factor. For this reason, we will turn to giving explicit consideration to the role of R and D.

## Location of R and D

High technology industry depends upon innovations which are derived from fundamental and applied research. It is natural, therefore, to suppose that the location of R and D facilities might have an influence on the location of companies engaged in high technology manufacturing. This hypothesis is explicitly tested by Markusen *et al.* (1986) for the United States, and found to be only partially verified. Defence spending, of which part will have been devoted to R and D, proved to be positively related to the location of high technology industries, but not university R and D facilities. Similarly, although industrial R and D in Britain is concentrated in the southern part of the country, the association which this may have with the location of manufacturing is not clear:

> Despite some early pioneering efforts, industrial R and D activity has not proved particularly amenable to geographical analysis and it has become increasingly apparent that a more thoroughgoing attempt by geographers to understand the *organizational processes* that are associated with R and D, together with an appreciation of R and D as part of the contemporary *political economy*, would both pay dividends.
>
> (Buswell *et al.*, 1985, p. 61)

Civil R and D occurs in three separate kinds of environment. If a single-plant firm has an R and D component, the location of the research and production sections is clearly joint and the decision to choose one location in preference to another may be primarily determined by either section. Whatever the location choice, however, there will be a close spatial and functional association of R and D and production. Multiplant firms, on the other hand, often have research facilities which are geographically separated from the production plants. The results of R and D work carried out in these circumstances are, in principle at least, available throughout the company – nationally or worldwide, as the case may be (Howells and Charles, 1989). For example, when Nissan announced in late 1989 that it was to establish two research centres in Britain, it was indicated that these would service their operations both in Britain and in Spain. Indeed, one of the anxieties which is commonly expressed about the growth of multi-plant companies is precisely the perception that R and D can be, and will be, divorced from the production centres. Finally, there are independent centres for research – mainly universities but including some research laboratories funded by research councils or in other ways. The results of their work are disseminated worldwide, in the form of journal and book publications, confer-ences and maybe patents. Although some contract and collaborative work is done with firms, this is not the primary outlet for basic research findings, and is not necessarily limited to work with local firms. There is no a priori reason why university R and D should be particularly associated with firms in the vicinity. Furthermore, as Oakey *et al.* (1988) emphasize, much technical know-how is transmitted internationally through licensing arrangements on the one hand, and through the subcontracting of technical development either to the suppliers of components or to customers.

For all of these reasons, it is inherently unlikely that civil R and D will be closely linked geographically with production, except in the case of one-plant firms and secondly if there are strong reasons for proximity between firms in the case of R and D done on a subcontract basis. In addition, to the extent that technical advances may give a firm a competitive advantage, knowledge of those advances will be treated as a closely guarded secret. For example, at least one of the pharmaceutical firms in Britain regularly checks its research facilities every two months for the presence of surveillance equipment which might have been planted by its rivals in order to gain access to its research findings. In any case, given the diverse sources of R and D advances, it is very likely that firms will have to draw upon resources outside the immediate locality. The force of this point is underlined by one estimate, that even a country of Britain's importance accounts for only about 5 per cent of the total R and D conducted worldwide (Barber and White, 1987, p. 33).

Confirmation may be found in the input-output format of information obtained by Patel and Soete (1987), summarized in Table 8.2. The majority of the industries classed as innovation-producing may be described as being high technology, producing substantially more innovations than they actually use – industries such as scientific and industrial instruments, electronic goods and synthetic resins. On the other hand, some of the innovation-using industries are also among those generally regarded, at least by some authors, as high tech-nology in character, for example, aerospace and petroleum. As least as impor-tant, though, is the importance accorded to innovations influenced by users in the case of the industries which produce innovations, and by suppliers in the

**Table 8.2**  Technology classification of innovative industries, United Kingdom, 1945–83[1]

| | Number of innovations | |
| --- | --- | --- |
| | Produced | Used |
| *Innovation producing industries* | | |
| Pervasive | 1814 | 453 |
| Localized, predominant use of own innovations | 233 | 104 |
| Localized, user-dependent | 471 | n.a.[2] |
| User-influenced | 641 | 309 |
| *Innovation using industries* | | |
| Significant own technological potential | 530 | 837 |
| Supplier-dependent industrial sectors | 114 | 912 |
| Supplier-dependent services | 6 | 116 |
| Supplier-influenced | 49 | 856 |
| | 3858 | 3587 |

[1]  More than 20 innovations, either developed within the industry or first used.
[2]  No data recorded.
*Source*: Patel and Soete, 1987, Table 8.2

case of industries which are predominantly users. The implication is that it is links between industries, rather than directly from civil research establishments (universities and others), which account for much of the innovation process in manufacturing.

In the light of these considerations, it is hardly surprising that *local* linkages, between high technology firms, with their suppliers/customers and with local civil research centres, are in fact rather unimportant. In his study of small firms in the San Francisco Bay area of California (including Silicon Valley), Scotland and the southeast of England, Oakey concluded as follows: 'These data give a clear overall impression that external technical information links are not profuse, nor are those that do exist of great significance to internal innovation in survey firms' (1985, p. 102). A vivid example of the international dimension to technological links is provided by the following account of the 'lost foam' process of casting, in which the hot metal vaporizes polystyrene moulds and takes their shape:

> It took the Department of Trade and Industry three years and two months to award Mr Dan Taylor just 6 per cent of the money he needed to put Britain ahead in high-technology car component manufacturing. It took the Japanese just three weeks and two days to arrive at his foundry with an offer, worth millions of pounds, to buy the entire business once they heard of the revolutionary casting process he has developed.
>
> (*The Times*, 8 May 1987)

Thus, as Oakey (1981, p. 119) found for the scientific and industrial instruments companies in Britain, the hypothesis that important locational ties exist with research establishments could be 'largely eliminated'. As for large multiplant companies, it appears that the location of R and D facilities depends on links to head office as well as to the main production site, both of which are apt to be in different locations, and that R and D activities are largely a 'self-contained' matter within the company, external links being limited (Howells, 1984).

If the measurable links between high technology firms within a relatively small area appear to be of limited importance, it is possible that informal contacts may be highly significant, notwithstanding the inhibitions which arise from the need to preserve technical secrets. It would appear that in the early days of Silicon Valley there was a heady atmosphere, as electronic and semi-conductor firms mushroomed in the vicinity of Stanford University:

> In an industry marked by rapid technological change, pervasive inter-company diffusion of ideas and severe competitive pressures which demanded always staying at the 'leading edge' of technology, there were clear benefits to spatial proximity and the clustering of firms. Small firms especially benefited, giving [sic] the frequency of product copying, second sourcing and pirating of information and personnel in the industry. There was also an unusually high degree of interaction between employees of rival firms in Santa Clara County. Many were close personal friends and had gone to school together or worked together in the past, and much information, brainstorming and gossip were exchanged over the telephone or at the local 'watering holes'.
>
> (Saxenian, 1985, pp. 29–30)

It would appear that, as the swashbuckling frontier days passed, the import-ance of these informal links has diminished. A rather similar picture has been painted of the much smaller agglomeration of high technology firms in and around Cambridge:

> The overall picture, then, is that a range of links between local firms and the University do exist in the high-technology sector, and that they are important for a substantial minority of firms. However, these links are informal and spontaneous, and have not really developed through com-prehensive strategic planning by University or local government authori-ties. They have also been greatly helped by the small size of the city and the ease with which close social networks have therefore evolved.
>
> (Keeble, 1989, p. 162).

In any case, with the passage of time small firms are apt to be taken over by larger companies, thereby becoming enmeshed in the protocols on which multiplant firms insist. These almost certainly reduce the scope for informal links between companies and research establishments (Crang and Martin, forthcoming).

*Science parks*

In the belief that close links between high technology firms, and between these firms and research centres such as universities, are important and can be deliberately fostered, science parks have become a fashionable way in which to promote development. For the present purpose, the question of main interest is whether experience with science parks throws useful light on the dynamic locational pattern of high technology industries.

Science parks are modelled on two strands of experience, the first being the successful development of general industrial estates in Britain, initiated before the Second World War at Slough, Team Valley and elsewhere, and much expanded after the war. The provision of serviced sites and, often, of factory premises built ahead of demand, to create an area of 'modern' (light) industrial employment proved to be a successful formula. Then in the early post-war period, largely spontaneous development of high technology industries occurred in the Route 128 area around Boston, while Stanford University initiated a green field development on its own campus which led to the establishment of the major complex known as Silicon Valley. This suggested that the provision of suitably located, serviced and landscaped sites in association with a university could trigger similar developments elsewhere (Christy and Ironside, 1987; Schamp, 1987). By 1960, about fifty science parks existed in the United States, a number which had risen to over 150 by 1984. The first one in Australia was established at Sydney in 1961 (Joseph, 1989). In Britain, it is usual to think of Cambridge (1973) and Heriot-Watt in Edinburgh (1974) and the first two, though Cranfield (1968) has some claims to this accolade (Currie, 1985); as late as 1980, there were still only three in existence, but by 1985 the number had risen to twenty and then to double that number before the end of the decade (Monck *et al.*, 1988, p. 92). West Germany's first science park was not established until 1984, in Berlin, but the growth since then has been rapid.

Most science parks are quite small. In the West German case, 10 to 20 enterprises is the norm, each employing between 4 to 10 people, i.e., a maximum of about 200 workers. As at January 1985, all of the British science parks had 20 firms or fewer with the exception of Cambridge, which had 46. This figure had risen to 65 firms in 1986, employing 2,170 people altogether. Within a 19-kilometre radius of the city, some 300 high technology firms employed a total of 13,000 people. Although the science park contributes about 17 per cent of local manufacturing high technology employment, its direct contribution to total employment is only 2 per cent (Keeble, 1989). Australia's biggest science park, at Sydney, had 90 firms at the end of the 1980s. Only a few science parks anywhere in the world are bigger than that: Sophia-Antipolis, near Nice in southern France, had reached 600 firms and 11,000 workers by the end of 1989. Both the Palo Alto science park at Stanford, California, and the Research Triangle Park in North Carolina are bigger, the latter giving direct employment to 30,000 people in 1989 (*New Scientist*, 1 April 1989, supplement, p. 3). It appears that most of the others in America are very much smaller, even though American science parks are generally 'on a much larger scale than those anywhere else in the world' (Monck *et al.*, 1988, p. 64), Sophia-Antipolis excepted.

The generally small size of science parks suggests that they cannot be the dominant factor in initiating or maintaining the local growth of high technology industry, and that the success of the Stanford initiative is more the exception than the rule. The Cambridge science park is much the largest and must successful in Britain, yet it got off to a slow start. The first firm moved in in 1973 but by the end of 1979 there were only 15 in occupation; the major growth was during the 1980s (Currie, 1985, p. 13). Indeed, the success of the Cambridge science park must be regarded very largely as a reflection of the boom in high technology employment in and around Cambridge, which began to be evident in the 1970s and has been dramatic since 1980, rather than as the initiator of this development.

Science parks are a recent phenomenon – more recent than high technology industry – and comparatively small scale. Furthermore, they have not in fact replicated the close interfirm and firm-university links which characterized the early days of Silicon Valley. A survey of over 400 firms in British science parks revealed that only 17 per cent were spin-offs from a local university (Monck *et al.*, 1988, p. 96). Writing of the Australian experience, Joseph (1989) also noted the lack of local linkages, concluding that the prime reason for locating in a science park is the prestige of a good address, a point confirmed by Currie: 'the majority of firms have, at present, only tenuous links with the university ... by far the most important reasons at present for firms locating in a park are related to property rather than academic contacts' (Currie, 1985, pp. 59 and 63). Crang and Martin (forthcoming) point to the process by which initially independent firms are acquired by other companies, while Joseph (1989) emphasizes the importance of inward investment, especially from overseas. Indeed, Monck *et al.* (1988) stress the key role in both Palo Alto and the Research Triangle Park of the presence of a single major firm in giving the park status and credibility – respectively Hewlett-Packard (and later, Fairchild), and IBM.

To date, science parks have been a comparatively small component in the growth of high technology industry. In addition, it appears that the supposed advantages of close proximity to university research, and of linkages between firms, have not in fact materialized. To the extent that they have been successful, that success is probably due to the availability of suitable premises in pleasant surroundings, to the general amenities which the locality can offer and the supply of graduates to be recruited, plus the prestige conferred by a science park address.

## Diffusion of technology

High technology industries account for about one-quarter of manufacturing employment and under 10 per cent of total employment. Important though high technology industries may be in generating product and process innovations, the future prosperity of regions will be dominated by the success, or otherwise, of 'low technology' industries, primary producers and the service occupations in adopting process and product developments. Basic commodities such as furniture and house fittings, clothing and personal transport will remain in demand, and some regions will continue to be highly dependent upon that demand. However, the relevant firms face actual or potential competition and to meet this they must remain competitive. This in turn implies keeping up with the stream of innovations, at the very least to ensure that production costs are held to the minimum. Thus, for all the attention given to the establishment and growth of new high technology industries, the majority of regions will continue to be highly dependent upon the success their firms have in keeping up with the changes in production systems relevant for their products.

At this general level, therefore, the possibility that there are regional differences in propensity to innovate must be taken seriously. If there are substantial, and systematic, differences between regions, this could be a much more important factor than the success or otherwise that regions have in attracting and/or fostering high technology activities. However, the potential importance of regional variations in innovation diffusion is not matched by the volume of literature available. This is somewhat odd, since quite a lot of attention has been given to international differences in the rate of technology adoption, as part of the wide enquiry into the reasons for differing levels of productivity and rates of productivity growth among nations (e.g., Davies, 1979; Nabseth and Ray (eds), 1974).

Notwithstanding the early pioneering work of agricultural economists in the inter-war period, geographical work on innovation diffusion is generally traced to Hägerstrand (1967), whose major study first appeared in Swedish in 1953. Two components of diffusion processes within a nation are recognized: contagious diffusion and hierarchical diffusion. The former, by analogy with contagious diseases, depends on close personal contact between people who already have adopted the innovation (become 'infected') and those who have not. This form of diffusion, highly dependent on the personal transmission of information, creates a wave of adoption which will spread radially from an initial point. Hierarchical diffusion is the transmission of an innovation through the urban system, generally downward from larger to smaller cities and from the more centrally located ones to those in remote, peripheral regions. The principle underlying this diffusion process is the volume of contact between cities – some positive function of their respective sizes and a negative function of the intervening distance. If hierarchical processes are important, then large and well-connected cities (regions) will be at a permanent advantage with respect to smaller and remoter ones, in that they will regularly adopt innovations sooner. If this happens in practice, there will be cumulative advantages for big cities in central locations. This process would be underpinned and supported if big cities are also the source of most innovations; there is some evidence to suggest that this was certainly the case in the nineteenth century (Pred, 1966).

Berry (1972) plotted the adoption of TV in the United States between 1940 and 1966 – as the opening of broadcasting stations and the spread of receiving sets among households. His results show an unmistakable diffusion from the larger to the smaller cities, the regularity of the progression becoming less clear cut as the innovation wave proceeded. Rather similar evidence was adduced by Robson (1973) for the spread of public services such as gas supplies in nineteenth-century Britain. However, as with Berry's study, these innovations are essentially consumption good innovations, with the larger markets being served before the smaller ones. This is in fact a different problem from the one of process or product innovation by firms, when those firms are competing on a national or international market rather than monopolistically supplying a local market.

From the evidence we have already reviewed concerning the nature of the innovation process in manufacturing, hierarchical diffusion is not a very plausible proposition, and there has been an evolution of ideas away from the hierarchical model to the realization that the situation is extremely complex. This is well illustrated by Pred's work spanning the period from 1966 to 1977. His 1966 study identified the larger cities, those growing rapidly and with a strong manufacturing base, as the major sources of innovation in the nineteenth century, innovation being measured by patents filed. The diffusion of those innovations, it was held, *ought* to be hierarchical, but little firm evidence could be marshalled in support of this expectation. Seven years later, Pred and Törnqvist (1973) published a theoretical study of information flows in systems of cities, in which they threw considerable doubt on the suitability of the hierarchical model in the context of industrial innovations. This was quickly followed by an immensely detailed study of intercity linkages in the United States (Pred, 1977). Taking multiplant companies, the locations of the firms' headquarters were mapped, along with the locations of the subsidiary plants. The resulting maps show a very complex pattern, in which cities both 'control' plants in other cities and themselves have plants which are 'controlled': the pattern does not display a neat hierarchical pattern, in which larger cities have the headquarters which 'control' plants in smaller cities. Pred is explicit in stating that the linkage patterns he had identified are incompatible with the idea of hierarchical diffusion of industrial innovations. In any case, a sharp change has been occurring in the source of patents filed in the United States. As recently as 1975, 65 per cent of new patents were American inventions; by 1985, the proportion had dropped to 55 per cent (*Economist*, 20 May 1989, p. 143). In other words, a large and growing proportion of patents filed in the United States originates outside that country's urban system, and will influence production through the complex web of corporate control which is not hierarchically based. The international transfer of technology is in any case a long-standing phenomenon, conducted both legally and illegally, a point graphically made by the *Economist*:

it is not so long ago that Europeans levelled against America the complaints that Americans now aim at Japan. At the Great Exhibition of 1851 British industrialists strolling around the Crystal Palace in London were horrified at the quality of the latest American guns. Unfair, they said. The dastardly colonials were using a British idea (a new lathe designed by a Mr Maudsley) to mill weapons with greater precision than the British. Then they had the cheek to sell Britons the fruits of their own basic research. Later, Britain made similar complaints about radar, penicillin, polyesters, computerised-tomography medical scanners – and much more.

(*Economist* 20 May 1989, p. 146)

Comparatively little literature has been published which directly addresses the spread of industrial innovations geographically within a country in recent years. That which there is does not use the urban hierarchy as a framework; the geographical pattern is analysed at the regional level. The single most thorough study of the intra-national diffusion of industrial innovations is probably that carried out by Thwaites *et al.* (1982) for Great Britain; summaries of this and related work have been published in several places (Gibbs and Edwards, 1983, 1985; Goddard *et al.*, 1986; Oakey *et al.*, 1982). The study took a number of innovations, mainly process innovations, thought to be of widespread significance in nine metalworking industries, ranging from agricultural machinery to the manufacture of pumps, valves and compressors. The study spanned the period from 1970 to 1981, using two spatial frameworks – the ten standard regions, and an alternative four-region classification of the country according to the level of regional policy assistance available. Five innovations were selected:

- computerized numerically controlled machine tools (CNCMT),
- computers in commercial use (CCU),
- computers in manufacturing and design (CMD),
- microprocessors in manufacturing processes (MMP),
- microprocessors in products (MP).

Table 8.3 provides a summary of the extent to which firms had adopted these innovations by 1981. This table is based on 1,234 usable responses to a postal questionnaire addressed to 4,923 establishments; we do not know whether the non-respondents' adoption patterns are similar to those who did make a usable return. The recorded regional differences in the level of adoption are quite small. However, the overall results might be affected by the variable mix of industries in the various regions; by the size of establishments; the availability of in-house R and D; and the corporate structure of firms – independent firms or multiplant enterprises. These questions were explored by means of logit

**Table 8.3**   Great Britain: adoption of new technology by 1981

|  | Percentage of respondents in each area having adopted | | | | | |
|---|---|---|---|---|---|---|
|  | CNCMT | CCU | CMD | MMP | MP | Average |
| Development areas | 21.3 | 60.9 | 28.2 | 10.1 | 13.9 | 26.9 |
| Intermediate areas | 25.4 | 70.3 | 31.1 | 12.6 | 20.2 | 31.9 |
| Non-assisted areas | 27.0 | 64.5 | 29.8 | 11.6 | 22.4 | 31.1 |
| Southeast | 24.5 | 62.6 | 22.8 | 11.4 | 22.7 | 28.8 |
| Great Britain | 24.8 | 64.3 | 28.3 | 11.3 | 20.1 | 29.8 |

*Note*: The abbreviations are defined in the text.
*Source*: Thwaites *et al.*, 1982, p. 11

**Table 8.4** Innovating firms, 1981–2 and 1985–6: percentage of respondent firms which had innovated in the preceding four years

| | | southeast England | | Scotland | | San Francisco Bay Area | |
|---|---|---|---|---|---|---|---|
| | | 1981–2 | 1985–6 | 1981–2 | 1985–6 | 1981–2 | 1985–6 |
| *Product innovation* | | | | | | | |
| Innovation | % | 78.3 | 77.8 | 63.0 | 59.5 | 85.0 | 90.5 |
| No innovation | % | 21.7 | 22.2 | 37.0 | 40.5 | 15.0 | 9.5 |
| Total | % | 100.0 | 100.0 | 100.0 | 100.0 | 100.0 | 100.0 |
| | number | 60 | 45 | 54 | 42 | 60 | 42 |
| *Process innovation* | | | | | | | |
| Innovation | % | 69.5 | 63.0 | 50.0 | 59.5 | 48.3 | 58.1 |
| No innovation | % | 30.5 | 37.0 | 50.0 | 40.5 | 51.7 | 41.9 |
| Total | % | 100.0 | 100.0 | 100.0 | 100.0 | 100.0 | 100.0 |
| | number | 59 | 46 | 54 | 42 | 60 | 43 |

*Source*: Oakey *et al.*, 1988, pp. 75 and 78

analysis, which is a form of multivariate analysis. This examination shows that bigger plants are more likely to adopt, as are plants with on-site R and D and plants which are part of a multiplant enterprise. However, the locational variables proved to be not statistically significant, and the signs for the four 'regions' shown in Table 8.3 were not consistent. The authors summarize their regional findings thus:

> The spatial variations in adoption are small but with the exception of computing in manufacturing and design the development areas perform below the national average in adoption terms. In terms of microprocessors in products the development areas perform significantly below the rest of the country. Within the development areas the poor performers are Wales and the North. The most adoptive regions are the East Midlands, West Midlands and Yorkshire and Humberside. Microprocessor adoption in products tends to be more prevalent in Southern Britain.
>
> (Thwaites *et al.*, 1982, p. 104)

Although this passage reflects the direction of the regional differences shown by the data, the smallness of those differences, the non-significance of the regional variables, and the large number of non-respondents suggests that the safer conclusion to draw is that regional innovation differentials are small and probably not significant.

However, a more recent study, but one based on a much smaller number of respondent firms, suggests that regional differences in innovation may in fact be significant. Table 8.4 summarizes data for the proportion of firms which introduced an innovation in the four-year period preceding 1981–2 and

1985–6. In respect of both product and process innovations, Scottish firms had been less innovative than those in the southeast of England in the period preceding both survey dates. Although the gap had widened for product innovations, it had narrowed in respect of process innovations. However, the apparent regional differences may reflect the mix of industries or other factors, which are not directly analysed in the study.

Rees *et al.* (1985, 1986) report the results of an enquiry in the United States which is similar in concept to that undertaken by Thwaites and others (1982) in Britain. The machinery industry, comprising industries such as farm machinery, machine tools and aircraft, was surveyed to elicit information on the extent to which eight innovations had been adopted. These innovations were mostly process innovations, such as numerical machine control devices and computers used for design. From the bivariate analyses which are detailed, it is clear that factors such as those identified by Thwaites *et al.* are also important in the United States – organizational status, size and age of plant and regional industrial mix. The truly independent regional effect proved to be small, even negligible, though with some suggestion that adoption rates were higher in the northeast than elsewhere, i.e., in the older industrial regions. This is associated with their evidence that older plants had been more innovative than newer plants. Hence: 'These results therefore show conclusive evidence that in a key part of the durable goods sector older manufacturing plants across the country have been rejuvenating themselves to remain competitive' (Rees *et al.*, 1986, p. 197).

Two reasonably reliable conclusions seem to be warranted at this stage. First, the diffusion of innovations through the population of manufacturing firms displays no evidence of the hierarchical patterns initially postulated. Second, that the distinction between 'old' and 'new' industrial regions is of little significance; there is some hint that innovation in Britain is more rapid in the south east than elsewhere, but similarly a suggestion that it may be more rapid in the north east of the United States. There is also some evidence that the distinction between small, single-plant firms and multiplant enterprises is important, the latter being more innovative than the former. Furthermore, there is some suggestion that small firms in remote locations are at a significant disadvantage relative both to small firms more centrally situated and also to branch plants in remote locations. Thus, concluding their review of the American machinery industry, Rees *et al.* note:

> For policy-makers interested in the nurturing of small business in par-
> ticular, this study shows that small firms nearer to the source of innovation
> are more likely to use leading edge technologies. Hence some attention
> may need to be given to encouraging the spread of these technologies to
> less innovating environments where multiplant firms have a clear advan-
> tage over single plant firms who suffer more from distance.
>
> (Rees *et al.*, 1986, p. 215)

An alternative source of evidence is provided by studies of individual industries (e.g., Chapman and Humphrys, 1987; Peck and Townsend, 1987; Schoen-

berger, 1987). These show that major innovations seem to be adopted widely throughout an industry within quite short periods of time, revealing no evidence that some regions are persistently leaders or laggards. Indeed, the Peck and Townsend study of Philips Electronics (UK) and Cadbury-Schweppes shows very clearly that large-scale reinvestment in older plants in major urban areas does occur.

An important aspect of the adoption of innovation is the need to train and retrain staff, so that the requisite skills are available to operate new machinery and new production routines. An extreme example of this need for training is provided by Fiat's plant at Cassio, southeast of Rome. This is one of the world's most highly automated car factories, where each worker has had at least 200 days' training (*Economist*, 21 May 1988, pp. 103–4). Training and retraining needs are usually much less arduous than that, but nevertheless significant. When a firm adopts an innovation and needs to retrain existing staff and/or train new recruits in the context of that innovation, three options may be considered. If the innovation depends on the installation of new machinery (a process innovation), the supplier of the equipment will normally provide some basic training as a part of the sales package. Such training will be available irrespective of the location of the firm adopting the innovation and generally irrespective of its size. Similarly, if a new product is being incorporated into an article, the supplier of the product has an interest in assisting its potential customer to use it and will often provide the necessary development and training inputs. Bigger firms, of course, will supplement these externally provided sources of retraining with their own formal schemes. These mechanisms imply that firms may not be at all closely linked with external agencies located in the immediate vicinity in order to keep up-to-date with recent innovations. In this sense, the availability of local training and retraining facilities may be unimportant, even for small firms.

Such a conclusion, however, relates explicitly to the new skills associated with current innovations. External training agencies have an important role to play in two ways (Thwaites *et al.*, 1982, p. 68). First, as skills become 'standardized', they become suitable for provision in general training establishments – manual skills such as plumbing and carpentry, secretarial skills, computer operation, etc. For all firms, especially the smaller ones, the availability of workers with these basic skills is likely to be a matter of considerable importance. In the second place, the quality of the basic schooling that children receive will play an important role in their future ability to accept further training and retraining. Viewed in the long run, it is possible that a region with particularly good schools and post-school training facilities will produce future workers with the habits and abilities which are conducive to the easy acceptance of change: 'The most important characteristic of a highly-skilled workforce in an age when technology is constantly changing is the ability to think clearly, to learn quickly, and to adapt to an ever-changing workplace and consumer environment' (Weiss, 1985, p. 90). If there are particular shortages of skills in a region, explicit steps may be necessary to remedy the situation.

We have returned to a theme that has been evident at several points in the preceding chapters, namely, the role of human skills and adaptibility. The regional dimension of this theme has its complement at the international level, and it is worth making the parallel quite explicit. Possibly the most authoritative study of international differences in productivity did not directly address the rate of innovation adoption, but yielded findings which are intimately related to that issue. Prais (1981) examined the performance of the United States, the United Kingdom and West Germany, on an industry-by-industry basis, in an attempt to identify the reasons for the lower level of productivity in Britain. Drawing together the threads of his exhaustive analysis, Prais singles out four issues for special mention. The first of these, plant size and market structure, proved to be unimportant – British plants being rather similar in size to those in the other countries. Nor could he find evidence that the supply of capital was particularly problematic in Britain, except for smaller companies, for which venture capital could be difficult to find. In this context, the more important problem was the difficulty British companies appeared to have in adjusting their manning to obtain the best use of capital investment. The third matter picked out for special mention is strikes. In all three countries, strikes were more common in large than in small plants, but sharply more so in Britain compared with the United States and West Germany. In addition, wildcat strikes were a more important feature of British strike activity than in the other two countries; wildcat strikes are especially disruptive because they are unpredictable. To the extent that some regions are more strike-prone than others, even if the strikes are concentrated in few establishments, a region will acquire an adverse image (see for example, Creigh, 1979). It is the fourth matter that Prais emphasizes particularly as explaining the productivity differences between Britain, the United States and West Germany, which is the question of technical training. Deficiencies in the supply of skills of all kinds have been noted since the beginning of the present century, but Britain's relatively poor standing persists. Prais notes that these deficiencies stem from the lack of adequate rewards for the possession of skills of all kinds:

> the main lesson of this study is that if Britain is to maintain its economic position in competition with the rest of an advancing world, the greatest priority must be attached to the improvement of the technical calibre of its workforce, and of the social and legal system which governs the way its members work together.
>
> (Prais, 1981, p. 272)

In one sense, this is a very encouraging finding, since it puts the focus on the man-made attributes of a society. What is true for the nation as a whole must also be true for the constituent regions. We have seen that regional differences in the rate of innovation appear to be small. To the extent that there are measurable differences, there is no inevitability that the pattern will be maintained for ever. The importance of the attributes of workers and entrepreneurs implies that the patterns can, and probably will, change.

## Small firms

It is widely accepted that since the early 1970s there has been a marked change in official attitudes to small firms. In the British case, the publication in 1971 of the report of a committee of enquiry on small firms (Bolton, 1971) marked the end of the belief that big is always efficient and good and led directly to the establishment of the Small Firms Division of the Department of Trade and Industry, of the Small Firms Advisory Bureau and other initiatives. Prior to the publication of the Bolton Report, small and medium-sized firms had been in long-term retreat in Britain and also in many other countries (Rothwell and Zegveld, 1982). However, between 1963 and 1978, the position in the United Kingdom stabilized and there was in fact a small advance in the relative importance of small enterprises in Britain, 'small' being those which employ fewer than 200 workers. Their share of manufacturing employment rose slightly, from 21.3 per cent in 1963 to 22.8 per cent in 1978, while their contribution to net output edged up from 18.0 per cent to 19.3 per cent (Curran and Stanworth, 1982, p. 8). The number of the very smallest establishments (10 employees or fewer), having declined from 1930, reached its minimum in 1963 and then, from the early 1970s, has shown very strong growth indeed (Keeble, 1986, p. 9).

Interest in small firms quickened in the recession of the late 1970s and early 1980s. The very large companies appeared to be at least as vulnerable as the smaller ones, and much anxiety was evinced at the labour shedding and rationalization undertaken by multiplant, usually multinational, firms. Observers were led to recognize that smaller enterprises might have merits, not just for survival during recession but for future growth as well. Consequently, the significance of indigenous growth came to the fore (see p. 91). For example, Hood and Young (1982), examining the retreat of multinationals in Scotland, noted that Scotland possesses significant powers for local initiatives and concluded:

> In the light of the problems discussed in this book, these powers have to be directed towards the expansion and replacement of the foreign-owned investment stock and towards indigenous industrial development ... the latter must inevitably receive top priority.
>
> (Hood and Young, 1982, p. 184)

International comparisons make it clear that the decline of the small firm sector up to the early 1970s had been sharper in the United Kingdom than in many other countries, including the United States. The small firm sector in the United States declined only marginally in relative significance between 1935 and 1963, and it continued to be more significant than in Britain. In sharp contrast, the role of small firms in Japanese manufacturing actually grew between 1962 and 1974, so that:

> In contrast to the US, therefore, where small businesses present a problem to the national economy because their numbers are low, in the sense that

more are thought desirable to stimulate and protect free competition as the cornerstone of the free enterprise system, in Japan small business problems are important because their numbers are relatively very high.

(Rothwell and Zegveld, 1982, p. 14)

The volume of literature which has been published since 1980 concerned with the birth of new enterprises and the success or otherwise of small firms has been considerable (e.g., Cross, 1981; Keeble and Wever (eds), 1986; Nijkamp *et al.*, 1988; *Regional Studies*, theme issue 18.3, 1984; Storey, 1985, 1987). The sea-change in attitudes toward small firms affected not just the United Kingdom but the countries of Western Europe and also the United States. On the other hand, critics have claimed that small firms cannot make a major contribution to the provision of jobs; that small firms provide 'low quality' jobs – low pay and poor conditions; and that policies designed to foster new and small firms will have their greatest impact where least needed (Storey, 1985, p. 2).

There is little doubt that, by certain criteria, the 'quality' of jobs in small enterprises is lower than in large ones. This has been clearly shown for the United States, on measures such as staff turnover (high), wage rate (low), pension provision (poor) and unionization (low). However, as Feldman (1986) notes, robust conclusions cannot be drawn without knowledge of the character-istics of the workers, and in any case a poor job may be better than none. To the extent that small firms are in high technology pursuits, the contrary evidence is that individuals value the freedom and challenge, thus making the concept of job 'quality' hard to specify in a meaningful way, and in any case many small firms in high technology industries seem to pay good wages.

We have already seen that 'indigenous' employment change has become more important than that associated with inwardly mobile firms (p. 91); 'indigenous' firms include the whole spectrum of new start-ups to large and long-established, mature enterprises. Some of the evidence in respect of new firms and employment change in existing small firms suggests that they make a very small, almost negligible, contribution to employment and output. In north east England in 1978, only 4 per cent of manufacturing jobs was provided by wholly new firms started up in the period from 1965 to 1978 (Storey, 1985, p. 39). An even smaller contribution was recorded in Scotland over the period from 1968 to 1977, new enterprises accounting for only 2.2 per cent of industrial jobs at the end of that period (Cross, 1981, p. 47). Less expected is the finding that in East Anglia, noted for its dynamic economy based on relatively small enterprises, the situation was not so very different. In 1981, employment in new firms established between 1971 and 1981 accounted for only 4.7 per cent of manufacturing employment (Gould and Keeble, 1984). The generally pessi-mistic conclusion implied by these findings was endorsed by the evidence for quite low rates of new firm formation up to about 1980 in the regions of Britain, generally below the level experienced in East Anglia (Gould and Keeble, 1984; Lloyd and Mason, 1984), and also by evidence for the southwest of England, showing that new enterprises had a neligible impact even up to 1985 (Gripaios and Herbert, 1987).

Other evidence points to the importance of new firms, most of which are small, in creating jobs at a time when larger enterprises have generally been shedding labour. Summarized by Rothwell and Zegveld (1982, pp. 121–2), over the period from 1969 to 1976 half of the gross new jobs in the United States were generated by new firms; the only manufacturing establishments to show job gains in the United States employed no more than 50 people. And, as of 1979, 23 per cent of manufacturing employment in the county of Leicester, England, was accounted for by independent new firms set up since 1947. These early signs that small firms were regaining some of the ground that had been lost have been confirmed by more recenet data.

The United States has experienced a small recovery in the number of self-employed, from 6.9 per cent of the non-farm workforce in 1975 to 7.4 per cent in 1986. Average real GNP per firm rose from $150,000 in 1947 to $245,000 in 1980 but has since fallen to $210,000 in 1987, a decline evident in all sectors except farming and the retail and wholesale sector. Over the period from 1976 to 1984, employment in establishments with fewer than 500 workers as a proportion of manufacturing employment rose from 51 to 53 per cent (*Economist*, 21 January 1989, p. 89). Similar but more striking trends are evident in Britain. The self-employed accounted for 8.0 per cent of employment in the United Kingdom in 1976 and 11.7 per cent in 1988. Meantime, as Table 8.5 shows, there has been a sharp drop in the relative importance of the larger establishments. Indeed, only the very smallest, employing between 1 and 10, experienced overall growth in the period. Note that these data are not directly comparable with the figures already given for the role of establishments employing fewer than 200 workers.

The resurgence of small businesses has been a general feature of Europe since the early 1970s (Keeble and Wever (eds), 1986). It occurred earliest in Britain and in Italy, and latest in Belgium and West Germany (Keeble, 1986, p. 8). The base from which this growth has occurred varies dramatically from one country to another. According to Korte (1986), in the early 1980s the proportion of

**Table 8.5** United Kingdom: number of employees in manufacturing, by size of manufacturing unit

| Size of unit, employees | % of employees | | No. of employees (000) | |
|---|---|---|---|---|
| | 1978 | 1988 | 1978 | 1988 |
| 1–10 | 4.2 | 6.8 | 287.8 | 324.2 |
| 11–19 | 3.9 | 5.0 | 266.7 | 236.9 |
| 20–49 | 7.6 | 10.7 | 526.5 | 511.3 |
| 50–99 | 8.8 | 10.8 | 606.6 | 512.0 |
| 100–199 | 12.9 | 13.9 | 827.9 | 663.2 |
| 200–499 | 19.6 | 20.7 | 1360.6 | 986.5 |
| 500–999 | 14.3 | 13.0 | 991.7 | 620.4 |
| 1000 and over | 29.7 | 19.0 | 2058.1 | 906.2 |
| | 100.1 | 99.9 | 6925.9 | 4760.8 |

*Source: Annual Abstract of Statistics*, 118, 1982; 125, 1989

**Table 8.6** Components of change in manufacturing employment, 1973–86 (000 jobs)

|  | United Kingdom | | Republic of Ireland |
|  | Northern Ireland | Leicester-shire | |
| --- | --- | --- | --- |
| Companies already established in 1973 | −90.8 | −55.5 | −92.5 |
| Companies established after 1973: | | | |
|     Inward investment | 6.9 | 3.9 | 43.1 |
|     Indigenous firms | 17.0 | 19.9 | 29.6 |
| Total change | −66.9 | −31.7 | −19.8 |

*Source*: Gudgin *et al.*, 1989, p. 18

manufacturing employment in establishments with between 1 and 9 employees exceeded 40 per cent in Sicily, parts of southern mainland Italy and much of Greece; in both countries, all regions were in excess of 20 per cent. Most of France, Belgium, the Netherlands and Denmark had proportions in the range from 10 to 20 per cent. In contrast the regions of the United Kindom were generally below 10 per cent, often below 5 per cent, and practically all West Germany was also below 5 per cent. Much the same geographical pattern is shown by establishments employing 10 to 49 workers. Clearly, the long decline of small firms was much more marked in Britain and Germany than elsewhere in Europe. Equally clearly, the recent revival of small businesses is not just a feature of those countries, such as Britain and Germany, where small manufacturing establishments had become an endangered species. Finally, there are clearly major problems in attempting to generalize experience from one country, or even one region in one country, to the situation in other nations. Nevertheless, throughout the western world the generality of the recent growth in the importance of small firms is not in doubt (Britton, 1989).

That the growth of new, small, businesses is having a material impact on regional employment in a country where small firms had been in decline is shown by Table 8.6. In both Northern Ireland and in Leicestershire, over the period from 1973 to 1986, manufacturing employment fell sharply. It was the firms already established in 1973 that contributed to this decline, which was partially offset by the establishment of new plants. Inward investment was comparatively unimportant, the major gain being in entirely new indigenous firms. In Northern Ireland, the gain of 17,000 new jobs in this category was in firms which, on average, employed only eight people (Gudgin *et al.*, 1989).

There is now little doubt that small firms are an important component of national and hence also of regional prosperity, or the lack thereof. The question of relevance to us, therefore, is whether experience tells us anything useful regarding the spatial dimension of the birth, growth and death of firms?

Conventional wisdom has it that most manufacturing firms are set up by people who have previously worked in small industrial enterprises. Accordingly, those regions which have a large stock of small firms will have the highest rates of new firm formation, which will put them at a considerable advantage over other regions. This conventional wisdom was challenged by Gould and Keeble (1984), who found that in East Anglia the major determinant of new-firm formation in manufacturing is the proportion of workers in non-manual occupations, and that if this variable is controlled for there is actually a negative association with employment in small firms. This conflict of testimony has been examined by Gudgin and Fothergill (1984). Part of the problem arises from the specification of the research hypothesis and the suitability of the data for testing it. Gudgin and Fothergill conclude, on the basis of the data they analyse, that indeed regions dominated by large plants do have low formation rates for new manufacturing enterprises but that non-manual employment is not a significant factor. They also confirm that new firm formation is biased toward rural areas and away from the big cities. Another analysis, however, shows that across the regions of the United Kingdom new VAT registrations are not related to plant size but that there is an association with the proportion of manual workers and the level of home ownership (Whittington, 1984): a high proportion of manual workers and a low level of home ownership militate

**Table 8.7**  Regional patterns of firm formation in Britain

| Region | 1919–1929 Gross[2] | 1930–1938 | Region | 1979–1987 Gross[3] | Net[4] | Stock of businesses per 000 workers[1] 1987 |
|---|---|---|---|---|---|---|
| London | 100 | 100 | Southeast | 100 | 100 | 100 |
| Southeast (excl. London) | 69 | 73 | (incl. London) | | | |
| North | 17 | 22 | North | 61 | 40 | 79 |
| Yorkshire | 40 | 28 | Yorkshire and Humberside | 74 | 46 | 91 |
| Northwest | 34 | 39 | Northwest | 74 | 31 | 89 |
| West Midlands | 29 | 37 | West Midlands | 77 | 61 | 91 |
| East Midlands | 20 | 35 | East Midlands | 82 | 76 | 97 |
| East | 41 | 27 | East Anglia | 97 | 103 | 106 |
| Southwest | 23 | 29 | Southwest | 100 | 100 | 127 |
| Wales | 16 | 18 | Wales | 82 | 62 | 127 |
| | | | Scotland | 59 | 49 | 83 |
| | | | Northern Ireland | 70 | 101 | 150 |

*Note*: Although many English regions have the same names, their geographical extent differs and direct comparison between the inter-war and post-war period is hazardous.
[1]  Extant VAT registrations per thousand employees plus self-employed in 1987.
[2]  Firm formation per thousand employees, in 1921 and 1931 respectively, London = 100.
[3]  VAT registrations per thousand employees plus self-employed, in 1981, southeast = 100.
[4]  VAT registrations minus de-registrations per thousand employees plus self-employed, in 1981, southeast = 100.
*Source*: Foreman-Peck, 1985, p. 412; *Regional Trends*, 24, 1989

against a high level of new firm formation. However, these two variables are themselves quite highly correlated.

To set this discussion in context, Table 8.7 compares evidence for firm formation rates in Britain in the inter-war period and more recently. The data used contain numerous problems, including the fact that the English regions have different boundaries in the two periods. Note that the figures are for all enterprises, not just manufacturing firms. This table makes it abundantly clear that the dominance of London and the southeast, which was clearly evident in the inter-war years, has been sharply reduced. Both East Anglia and the south west now have new-firm formation rates comparable to the southeast, and the overall regional differentials are now much smaller than formerly. This is true in regions which have not benefited from regional policy as well as those which have. However, it is true that the net formation rate is lower in the assisted regions – and especially the north, north west and Yorkshire and Humberside – but with the notable exception of Northern Ireland. However, if we take the number of VAT-registered businesses in 1987 relative to the number of employees and self-employed in that year, the regional variation in the stock of businesses is much smaller than is the variation in formation rates. Furthermore, the 1987 stock figures are not wholly compatible with the gross and net new-firm formation rates. For example, from the firm formation data one would not expect the north to have four-fifths of the proportionate stock of the southeast, and the other assisted regions to have an even higher proportion. Nor would one expect Wales and Northern Ireland to be way in excess of the southeast.

Table 8.7 suggests that analyses of individual regions, and analyses carried out over fairly short periods of time, will yield conclusions which at best are partial and at worst are misleading. It further suggests that the geography of new firms, and of small firms generally, is not fixed and immutable, and that regions which have been hard hit by the decline of staple industries can fight back.

The difficulty that there is in reconciling the evidence for regional variations in new firm formation and possible causal factors probably resides in the variety of reasons which lead to the establishment of new enterprises (Keeble, 1990b). In rapidly growing industries, characterized by small establishments and a high rate of innovation, it is common for employees to leave their company to set up their own business. This is a pattern which characterizes many high technology industries, with Silicon Valley as a type case. On the other hand, regions experiencing serious unemployment and high redundancy rates may also experience a high birth rate of new firms, possibly as a matter of desperation, and possibly with poorer prospects than firms created by the first process, but nevertheless changing the structure of employment by size of enterprise. However, there is some evidence which suggests that the take-up of official schemes to help redundant and unemployed workers may be greater where the need is in fact less pressing (Mason and Harrison, 1986, 1989). Then thirdly, one of the rationalization strategies followed by large firms in the face of intense competition and changing technology has been to put out work which formerly was done in-house. This may not lead directly to the establishment of new enterprises, but will serve to shift employment toward smaller units (Shutt and Whittington, 1987). Finally, many larger firms are now taking a more encouraging and more helpful attitude toward the existence and growth of smaller enterprises in their locality, whereas formerly they discouraged such development because they feared competition in the local labour market.

Whereas in the early 1980s there was still considerable scepticism regarding the contribution which small firms might make toward regional employment, especially in regions traditionally dominated by large employers and/or branch plants, that scepticism has abated markedly in recent years. Small firms are now generally accepted as an important component of the employment landscape. There are marked regional differences in both gross and net new firm formation, as shown by numerous studies, most of which span comparatively short periods of time. Viewed over several decades, however, it is clear that sharp changes in regional ranking have occurred, and that the regional differentials have narrowed considerably, not least in Britain. Furthermore, if we take the stock of firms in1987 in the regions of Britain, regional differences are much smaller than is implied by the evidence in gross and net changes measured over short periods in the past. The cumulative process, in which regions with small firms beget many new small firms, does not seem to be working as powerfully as had been expected. The implication is that, even if it takes time, it is indeed possible for small firms to thrive in regions which, on past experience, might seem inhospitable to them.

With some confidence, we may assert that small firms are more dependent on their external environment than are large firms. In particular, the availability of cheap and suitably small premises in which to start up can be crucial. One of the problems of tidy town and country planning is its dislike of obsolete premises and non-conforming uses, with the result thay many suitable starter premises have become unavailable. The wheel has been reinvented by the provision of small start-up units by local authorities, by English Estates and by agencies such as the Rural Development Commission (Perry and Chalkley, 1985). Small firms are not usually able to mount training schemes for their workers and therefore must rely on the general calibre of workers in the area and the availability of specialist training facilities for particular skills. In addition, small firms will need to rely on external advisory services for information which big firms provide for themselves – accountancy, marketing, technical, etc. The local infrastructure of these services will be an important factor in the ability of small firms to start and then to thrive (Britton, 1989).

## Conclusion

The evidence reviewed in this chapter leads to one very clear conclusion. There is no pattern of 'inevitable' location of modern manufacturing in particular areas and therefore there is no clear-cut set of causations to be identified. The undoubted success of some newly favoured regions is matched by the ability of many older industrial areas to attract and nurture the new kinds of enterprise. It is clear that transport costs on raw materials, fuels and components, and on marketing finished products, are small, possibly negligible. These traditional location variables can be discounted. Attention focuses on the attributes of the skilled workers and entrepreneurs, on the qualities that they expect of their living and working environment, and the possibility that interfirm linkages in research and product development may be important. In practice, this last factor seems to be less important than one might think, partly because industrial R and D is a matter of great commercial importance which is therefore apt to be kept safely out of the ken of rivals.

In sum, it appears that the single consideration of most importance is the quality of the environment that a region can offer to attract and to retain the skilled development and production workers it needs. Sunshine, mountains, lakes and the sea all play a part in that environment, but these natural attributes are not all found in one place, and human beings differ in their preferences among these options. More important, it appears that the man-made qualities of cities, towns and rural areas are at least as important. These include the quality of theatres, libraries, schools and recreation facilities, as well as the degree of congestion, access to motorways, to airports and to railway systems. No place has a monopoly of these good things of life; everywhere there is some trade-off, and the utility functions of individuals do vary.

The implication seems to be that the success which cities and regions have in attracting inward investment and in fostering indigenous enterprise really does depend, in part at least, upon the nature of the environment that can be created. To this extent, regions do have some control over their destiny. While it would be wrong to suggest that every Massachusetts can become a miracle (Harrison, 1984; Lampe (ed.), 1988), it would be equally wrong to dismiss attempts at local/regional self-help as being a waste of time. In the next chapter, therefore, we will examine the nature and success of such endeavours.

# CHAPTER 9

# *Local initiative*

As Soichiro Honda once observed of his company: we do not make something because the demand, the market, is there. With our technology we can create the demand, we can create the market.

(Friedman, 1983, p. 368)

The triumph of the twentieth century is that it has purged itself of certainty.

(*Economist*, 12 August 1989, p. 16)

America emerged from the Second World War as the undisputed economic giant of the western (non-communist) world. That position has been steadily eroded, by the resurrection of Western Europe, and especially West Germany, by the advent of Japan as a powerful exporter, and by the growth of several other significant manufacturers, such as Brazil, South Korea and Taiwan. It is unlikely that the world will ever again be so dominated economically and technically as it was by Britain in the nineteenth century or the United States for much of the twentieth. As the world's industrial base has expanded, the sources of innovation have multiplied. If any country stands still, it will find that competitors will enter its markets and offer cheaper or better products, and that it will be debarred from new markets as they open up. All the major industrial nations have the same basic problem, of maintaining and, if possible, enhancing, their competitiveness.

The forward-looking and imaginative outlook expressed by Soichiro Honda is essential. It is also, of course, a colloquial expression of Say's Law in the context of an open national economy, providing one summary of the arguments in the two preceding chapters. On the other hand, many brilliant innovations fail to generate the market which was expected – supersonic travel by Concorde and Hovercraft services being but two examples. If we now know that Keynesian demand management cannot deliver all that it promised, we know also that we do not, nor probably ever will, fully understand how to manage and control the pressures of adaptation and change. Countries which aim to use best practice technology and to take part in generating both product and process innovations necessarily find themselves in relatively uncharted frontier territory. Uncertainty is necessarily an important ingredient of economic frontier life. Only if a country is catching up, or content to remain a follower, can the way forward be charted with some confidence, as with the much admired 'indicative planning' used in France after 1945 and the even more effective activities of the Ministry of International Trade and Industry (MITI) in Japan. However, successful intervention to catch up creates the conditions in which it becomes increasingly difficult to continue with successful national planning precisely because the future is so difficult to predict. Nevertheless, in the circumstances of each country, certain policies are more likely to be helpful than others in fostering the long-term capability to maintain competitiveness; the problem is deciding which policies fit the particular circumstances. The same problem is evident at the regional level. Is one, as Hamlet mused:

> ... to suffer
> The slings and arrows of outrageous fortune,
> Or to take arms against a sea of troubles

in the hope of bringing them to an end?

The preceding chapters lead us to the following summary conclusions in respect of the advanced industrial nations:

(1)    Just as no man is an island, 'entire of it self' (John Donne, *Devotions*), so is no region separate and unconnected with the world.
(2)    Consequently, every region is affected by decisions elsewhere.
(3)    But every region is an 'elsewhere' for all other regions, so decisions taken within its bounds will have some impact on the prosperity of its inhabitants and those of other regions.
(4)    Although corporate structure is an important factor in modern economic geography, the importance of the location of *decision making* should not blind us to the equal importance of the determinants which lead to particular *decisions*. Why are some regions preferred, by single-plant or multiplant firms, and other spurned?
(5)    To the extent that regions have some powers of governance, exercised in some degree co-operatively or antagonistically with central government and other agencies, there is some scope for regional initiative which will create circumstances more favourable for employment prospects than would otherwise be the case. This is true for indigenous firms, actual or potential, and also for multiplant firms which may be persuaded to choose the region for inward investment and/or the expansion of existing plants.
(6)    For most modern industry, the resource of greatest significance is people – their skills, aptitudes, ability to adapt – and where people wish to live. Raw materials and fuels are generally not important, though agglomeration economies may be significant, and man-made facilities may compensate for less sun or access to outdoor recreation areas.

Many industrial cities have experienced severe economic recession, accompanied by loss of jobs and/or residents. But no major city in the industrialized world has been abandoned, to become a ghost town comparable to those which can be found in North America, Australia and elsewhere – relics of mining, lumbering or transport activities dating from the early days of European settlement. Even Benton Harbour, on the shores of Lake Michigan, chosen by *Money* magazine as the least desirable place to live in the United States, is still home to 75,000 people. Its economy was devastated in the 1970s by the closure of metalworking plants. The nadir was reached in 1985, since when new businesses have begun to establish themselves and the total number of jobs has begun to grow again, partly in response to local initiatives (plus the cheapness of premises and labour). Even though unemployment still exceeds 30 per cent, hope for the future has been reborn and the town will almost certainly survive and improve (*Economist*, 16 December 1989, p. 40).

Since development never has been, nor ever will be, uniformly spread among existing settlements, over any given period of time some regions will gain while others lose. The question, therefore, is how do regions adjust to change, especially that change which is adverse? It does appear that these adjustment processes occur with greater facility in some countries than in others, as is clearly shown by a recent OECD (1989) study. This analysed regional unemployment patterns in a number of countries, to see how far the geography of unemployment remained constant from one date to another. High correlation coefficients between pairs of years indicate consistency in the pattern, whereas low coefficients show that there has been a major geographical shift in the incidence of joblessness. For the purpose of such comparisons over time, the fact that the regional units differ from one country to another does not introduce serious biases.

Table 9.1 summarizes the results of this analysis. Three countries display a strong continuity in the patterns of unemployment over time, namely Finland, Japan and the United Kingdom. The problem regions of 1960 have remained the problem regions to 1987. For four countries (Canada, France and Italy certainly, and possibly Sweden) the geography of unemployment changed in the earlier years but more recently the early pattern has been re-established. Germany is distinct. In this case, the geography of unemployment has changed more or less continuously over the entire period, such that there is little correlation between 1987 and 1960. The United States is different again, having experienced moderate stability in the spatial pattern of unemployment up to 1975, whereas the correlation between 1975 and 1987 is the only negative value in the entire table. A *partial* explanation for these national differences may lie in the differing rates of migration: typically, in Western Europe between 1.5 and 2.0 per cent of the population moves between regions each year, compared with 2–3 per cent in Australia and nearly 4 per cent in the USA (OECD, 1989, p. 96).

The evidence of Table 9.1 may be interpreted in at least two different ways. It could be argued that the lower the correlation between unemployment patterns at different dates the more successful is the economy in adjusting to change. On this interpretation, the United States and West Germany have been the most successful of the ten nations in making spatial adjustments; Finland, Japan and the United Kingdom have been the least. Alternatively, it may be argued that the apparent adjustment success represents no more than a beggar-my-neighbour, zero-sum game, in which one region's success is bought at the expense of another's increase in unemployment. One conclusion is inescapable,

**Table 9.1** Correlation between recent and historical patterns of unemployment rates by region[1]

| | Australia | Canada | Finland | France[2] | Germany | Italy | Japan | Sweden | United Kingdom | United States |
|---|---|---|---|---|---|---|---|---|---|---|
| **Correlations with 1975** | | | | | | | | | | |
| 1960 | — | — | — | 0.92 | 0.48 | 0.81 | 0.86 | — | 0.80 | 0.53 |
| 1965 | 0.30[3] | 0.89[3] | — | 0.89 | 0.76 | 0.34 | 0.92 | — | 0.82 | 0.32 |
| 1970 | 0.26 | 0.68 | 0.94[4] | — | 0.91 | 0.95 | 0.93 | — | 0.90 | 0.49 |
| 1975 | 1.00 | 1.00 | 1.00 | 1.00 | 1.00 | 1.00 | 1.00 | 1.00 | 1.00 | 1.00 |
| 1980 | 0.07 | 0.86 | 0.92 | 0.53 | 0.71 | 0.94 | 0.97 | 0.85[5] | 0.97 | 0.52 |
| **Correlations with 1987[6]** | | | | | | | | | | |
| 1960 | — | — | — | — | 0.31 | 0.67 | — | — | 0.74 | 0.34 |
| 1965 | 0.61 | 0.87 | — | — | 0.21 | 0.19 | — | — | 0.75 | 0.19 |
| 1970 | 0.74 | 0.82 | 0.93 | — | 0.40 | 0.89 | — | — | 0.86 | 0.29 |
| 1975 | 0.07 | 0.67 | 0.88 | 0.46 | 0.54 | 0.87 | 0.91 | 0.68[7] | 0.92 | −0.21 |
| 1980 | 0.30 | 0.52 | 0.94 | 0.93 | 0.92 | 0.91 | 0.95 | 0.82 | 0.98 | 0.32 |
| 1985 | 0.95 | 0.98 | 0.99 | 0.98 | 0.99 | 0.90 | 0.95 | 0.91 | 0.98 | 0.83 |
| 1987 | 1.00 | 1.00 | 1.00 | 1.00 | 1.00 | 1.00 | 1.00 | 1.00 | 1.00 | 1.00 |

[1] All correlation coefficients are calculated using 1980 labour force as weights.
[2] For France correlations with 1975 related to 1962, 1968, 1975 and 1982.
[3] Correlation of 1966 with 1975.
[4] Correlation of 1971 with 1975.
[5] For Sweden correlation with 1976.
[6] For Germany correlations with 1986.
[7] Correlation of 1976 with 1987.
*Source:* OECD, 1989, p. 108

however. If we focus on those countries that have shown considerable stability in the geography of unemployment, then it is clear that over an entire quarter of a century there has been a failure of the adjustment processes. The processes postulated by the neo-classical school of thought have been inadequate. Where Keynesian regional policies have been employed, as in Britain, these too have not achieved the desired equalization of unemployment. Nor have the more recent supply-side policies made much impression on the long-standing regional differences – indeed, some argue that they have been exacerbated.

As we have seen in earlier chapters, several reasons may be adduced for this failure, one being that not enough attention has been given to local/regional supply-side issues (Chapter 7). One may argue that the long-term national interest is best served by allowing market mechanisms to operate, in which case local/regional attempts to alter the course of events are not desirable. That argument can only be valid if three assumptions are themselves correct. The first assumption is that locational determinants are 'objectively' given, arising from *immutable* circumstances of resource base and location. This is patently not true. Second, we cannot assume that all firms and individuals have the same priorities, needs and desires. Third, even if locational determinants were fixed and given, and even if all economic actors had the same locational preferences, were they all to exercise their freedom and locate in the same area they would spoil the very thing for which they had come. An extreme example is presented by Los Angeles. The famous Californian sun is obscured by smog; the freedom to occupy spacious plots has been offset by grid-lock conditions on the freeway system; and water resources have been exploited to the point at which additional supplies will be extremely costly to harness. No doubt all these problems can be solved, but the idyll now has a tarnished air. Also in California, serious groundwater pollution mars the success of Silicon Valley, threatening to limit further development. The remarkable explosion of house prices in the southern half of Britain in the mid-1980s has created conditions which make this a less desirable region in which to live (Breheny and Congdon (eds), 1989) and in fact in 1989 house prices actually fell while further north they rose. Much derided though the neo-classical supply-demand system of thought may have been, it does set limits to the degree of spatial concentration that is likely to occur, and it does offer hope for those regions which presently are losing out.

How, then, are these regions 'to take arms', to become more desirable places in which to live and work? This question has attracted a great deal of attention since the late 1970s, in Britain, Europe and the United States (Blakely, 1989; Chandler and Lawless, 1985; Donnison and Middleton (eds), 1987; Gibbs (ed.), 1989; Johnson and Cochrane, 1981; Keating and Boyle, 1986; Krebs and Bennett, 1989; Lever and Moore (eds), 1986; Mawson and Miller, 1983; Morison, 1987; Southwick, 1986; Webman, 1982; Wilmers and Bourdillon, 1985). This flood of publications reflects the realization that Keynesian policies are insufficient, possibly misconceived, in a period of higher unemployment than had previously been experienced. It reflects, also, the rapid burgeoning of local initiatives in the face of the explicit withdrawal of central governments from many forms of intervention, whereby localities and regions have been driven to fend for themselves or face the prospect of an uncontrolled downward spiral in their

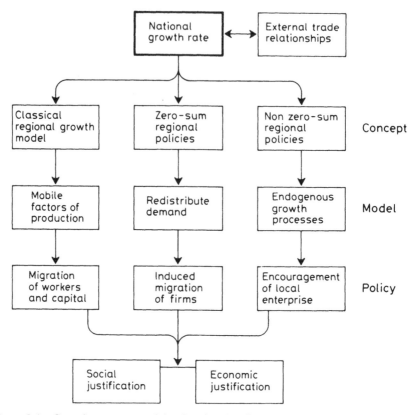

**Figure 9.1** Complementary models of regional policy

fortunes. Consequently, much of the literature describes what has been happening on the ground, in particular places and by particular agencies. Schematically, these initiatives can be portrayed as in the right-hand side of Figure 9.1, not as an alternative to the neo-classical or Keynesian models, but as a complement.

The plethora of initiatives which have been essayed makes formal assessment of their success or failure difficult or impossible. In addition, because they take place in a national and international context, whatever is done at the level of the locality or region is inextricably mixed up with wider trends and policy initiatives, which may promote or negate that which is being attempted at the local level. There is, therefore, no certainty about the efficacy or otherwise of the initiatives which we are about to discuss.

## Inward investment

Any regional development agency worth the name is interested in attracting investment which might otherwise go elsewhere. From the viewpoint of an individual region, to win such a competition will bring jobs and income that will boost local employment and income, even if the regional multipliers are in fact quite small. Viewed across a number of regions, however, the competition for inward investment is a zero-sum game; the investment is presumably going to occur anyway and the only issue is where it will be located. The process inevitably creates competition between the workers in different locations, and between units of local government and/or development agencies. Rees (1986, p. 2) notes the case of two large firms in the United States, both of which announced plans to build a major new facility and invited 'bids' from localities. In the event, the Microelectronics and Computer Research Corporation selected Austin, Texas, from among sixty prospects (Vaughan and Pollard, 1986, p. 268), while General Motors located its Saturn division near Nashville, Tennessee. Similar competitions are commonplace with international investment decisions. As part of its bid to obtain a £200 million engine plant planned by its parent company, Vauxhall obtained important concessions from the unions to facilitate flexible operations at Ellesmere Port, Cheshire; some of the working practices agreed in principle will be envied by Ford, Rover and other large companies (*Independent*, 4 December 1989). When, also in 1989, Toyota chose Derby in preference to other British locations, an important consideration was the suitability of the land which could be made available – a disused aerodrome. The same kind of competition can also be set up in the context of plant closure: International Harvester announced that it would close one of two plants – Fort Wayne (Indiana) or Springfield (Ohio) – which one it would be depending on the nature of the deal it could thrash out with workers and the local administrations (McKenzie, 1984).

Inward investment can be stimulated in three ways, which are not mutually exclusive and which are usually employed in conjunction. Promotional information, presenting companies with information about an area and the opportunities it provides, is an essential component. If companies are unaware of a region, they cannot consider it; and if they have an unfavourable image, they will look elsewhere. In the second place, initial, one-off assistance may be on offer – the most usual being financial incentives of one kind or another. The most common form of such assistance in Europe is central government grants and grants from the European Community (Yuill and Allen, 1980-), whereas in the United States it is the states and urban jurisdictions which are more prominent. In addition, though, the availability of land can be crucial and in this sphere local authorities, even in a centralized state such as Britain, can have an important role to play – either because they themselves own the land in question, or because they control decisions over land use categories for planning purposes. The third main form of attraction relates to the situation when a company has set up and is in business. Although British local authorities do not normally have the power to exempt a company from the payment of rates, wider discretion exists in some American states and in West Germany to grant medium-term remission of or exemption from taxes. Much more important, probably, is the quality of services offered locally, either directly to the firm itself or those that are relevant to employees – schools, libraries, recreation facilities, law and order, etc. High local taxes will be acceptable if it is perceived that employers and employees get value for their money.

Inward investment strategy suffers from two major limitations. In the first place, as we have already seen in Chapter 5, the supply of mobile investment is too small to solve the needs of all regions; indeed, in many cases inward investment represents quite a small proportion of the gross job gains. For this reason alone, sole reliance cannot be placed on this approach. The second reason is the inevitable tendency for immediate and tangible inducements to be increased, as regions bid to win desirable investments. Such upward bidding is more likely to benefit the company and its shareholders (through an enhancement of profits) than the customers (on account of lower prices), and the outlay by the regional or local administration may mop up most if not all of the benefit which that area might expect to receive. It is for this reason that the European Community seeks, with but partial success, to set ceilings to the amount of direct, or 'transparent', aid which is offered to firms. In respect of the car industry, for example, any investment worth more than $13.5 million and which includes an element of subsidy has to be cleared with Brussels before it can proceed. The magnitude of the harmonization problem is illustrated by Table 9.2, showing the total level of state aid to industry (not just regional incentives) in ten EC countries. In preparation for 1992, the Community countries are seeking to standardize assistance, so as to ensure truly 'fair' competition. In contrast, the federal government of the United States has no power to regulate the level of state and local assistance which may be offered; limitation is imposed by the assessments which states and local authorities make regarding the size of benefits to offset against expenditures.

## Local improvement or restructuring

For these twin reasons, regional jurisdictions have become increasingly interested in the need to foster and encourage existing businesses and the birth of indigenous new ones. To foster local business requires a pattern of intervention which differs from that needed to attract inward investment. Promotion of the region is not relevant, unless it is helpful in encouraging the sales of local firms. In general, though, manufactured goods are not identified with particular regions in a country, except for manufactured foodstuffs such as cheese, and drinks like wine and whisky. Competitive financial or other assistance is also not relevant for the start-up of local, indigenous, firms, but may be important if a plant is threatened with closure. In respect of indigenous enterprise, therefore, main emphasis must be placed on identifying the needs of firms, whether the availability of starter premises, advisory services, skilled labour, etc. In addition, thought must be given to what a locality/region needs to offer as the general context for the operation of firms and as a place in which workers live, bring up a family and follow recreation pursuits. All of these matters also have a bearing on the attractiveness of an area for inward investment.

It is easier to specify the kinds of things that may be done by a region to help itself than it is successfully to implement practical policies. However, the

**Table 9.2** Western Europe: state aid to industry as percentage of GDP, 1981–6 average

| | |
|---|---|
| Luxembourg | 6.0 |
| Italy | 5.7 |
| Eire | 5.3 |
| Belgium | 4.1 |
| France | 2.7 |
| West Germany | 2.5 |
| Greece | 2.5 |
| United Kingdom | 1.8 |
| Netherlands | 1.5 |
| Denmark | 1.3 |

*Source*: *Economist*, 18 November 1989, p. 127
*Note*: Manufacturing industry accounts for just under one-half of total aid, on average.

evidence reviewed in Chapter 8 shows how diverse are the locations which seem to be suitable for modern industry, and provides us with some clues concerning the common features that these locations possess, or which have been created.

The first issue to consider is the appropriate geographical scale at which action may be taken, for problems whose manifestation may be regional in extent or as localized as the inner areas of major cities. We may pose the question in the form: is it for national agencies to act, or is it a matter for regional or local administrations? The answer to that question depends, at least in part, on the way that relations between these levels of government are articulated in different countries, as is clear from a study of three European cities:

> The ratio between these three levels and the effectiveness of the intervention varies widely. In Clydeside the regional dimension, through the SDA [Scottish Development Agency], has come to dominate; in Cologne, city government appears the strongest force; in Turin, national policies and processes had the most impact.
>
> (Lever, 1989, p. 229)

However, the *apparent* differences between nations are often greater than the reality in practice. For example, Britain is a centralized state in which the existence and powers of local authorities depend on legislation enacted in Parliament. In contrast, local authorities in federal West Germany have their position enshrined in the constitution. In practice, the *Länder* and *Bund* governments in West Germany do supervise and regulate the local authorities, while British local authorities seek and find ways in which to escape from the control exercised by the centre. Consequently, the real differences between the two countries are smaller than the apparent ones and, at least up to the early 1980s, had been declining (Johnson and Cochrane, 1981; see also Krebs and Bennett, 1989). Both countries differ from the United States, in that the competence of federal and state governments to intervene in city affairs is generally considerably less than would be regarded as normal in Europe; they also differ from France, which is perhaps the most centralized country of all, central government playing a significant part in the actual implementation of redevelopment programmes (Webman, 1982).

The situation in Britain during the last decade has been curious. In a number of important respects, the powers of local authorities have been reduced and the impression created that central government is hostile to local corporate initiative, in pursuit of the 'enterprise culture' (MacGregor, 1988). The Audit Commission (1989) acknowledges both points, but shows that there has nevertheless been a continuing and expanding role for local authorities. The most obvious 'attacks' on local authorities have been the abolition of the most powerful ones in England (the Greater London Council and the Metropolitan Counties), the imposition of expenditure ceilings, the partial abrogation of local authority powers in the creation of Enterprise Zones (EZ), and the total abrogation in the case of Urban Development Corporations (UDC). However, the number of EZs and UDCs created has in fact been quite small. Much less widely remarked are the implications of the Uniform Business Rate, introduced at the same time as domestic rates were replaced by the Poll Tax (1989 in Scotland, 1990 in England and Wales). In the past, local authorities have had the power to set their own business rate poundage (the rate to be levied per £ of rateable value). Henceforth, the rate poundage will be levied at a uniform level within each of the three countries comprising Great Britain. The rates so collected will be aggregated for each country and then redistributed to local authorities in proportion to their population. As a result, there ceases to be the direct incentive that used to exist for local authorities to increase the level of business activity in their area; an increase in business rateable value will no longer result in a larger revenue flow for that authority, other than through rents and other trading income and a higher Poll Tax revenue if there is an associated increase in population.

Despite these specific developments and the associated impression that central government is anxious to limit the scope of local authorities' action, there has in fact been a considerable increase in their involvement in economic regeneration. In the decade following 1978–9, the number of staff employed by local authorities in economic development approximately doubled (Sellgren, 1989, p. 241), while the number of local authorities in England and Wales having a Local Enterprise Agency rose from about 10 in 1979 to over 300 in 1989 (Audit Commission, 1989, p. 61). Even so, the financial resources which local authorities apply *directly* to economic development is comparatively modest. Less than 1 per cent of local authority expenditure (excluding housing and debt charges) is devoted to economic development. Under the power conferred by Section 137 of the 1972 Local Government Act to raise a tuppence rate, authorities in England and Wales spent only £84 million on economic development in 1984–5; in contrast, central government provided £1,868 million in 1988–9 for programmes relevant to urban regeneration (Audit Commission, 1989). On the other hand, most expenditure by local authorities has an impact on the quality of an area in which to live and work, even if this impact is indirect. Total expenditure (capital and revenue) by all local authorities in the United Kingdom amounted to £50 billion in 1988-9, just over 10 per cent of GDP (Central Statistical Office, 1989). Despite the considerable control exercised by central government, and despite the fact that many services provided by local authorities are in response to duties laid on them by Parliament, considerable discretion remains to local authorities. The wisdom and efficiency with which that discretion is exercised will have a material and cumulative impact on the quality of localities as places in which to live and work.

Within the United States, the federal government plays a comparatively small part in economic development at the level of state and city jurisdictions, and the latter have considerably greater autonomy than is the case in Europe – a greater range of responsibilities and also greater freedom to raise finance through local taxes and/or borrowing. In the American context, therefore, it is plausible to regard each local jurisdiction as providing a bundle of services at a given quality for a cost to the individual measured by his tax contribution. According to Tiebout (1956), each household will choose to live in that jurisdiction whose bundle of services most nearly accords with that household's desires, given the level of income and the tax to be paid. This concept seems to be much less applicable in Europe, a fact which underlines the greater autonomy of local government in America. However, as Hepple (1989) points out, the Tiebout formulation provides a major part of the intellectual case for the introduction of the Poll Tax in Britain.

Whatever the balance of power and authority between local government, regional government and the central administration, effective action depends on sensible co-operation between agencies at all three levels. Such co-operation is not always forthcoming, or not as fully as should be the case (Audit Commission, 1989). We will accept that no system can be perfect and that neither local government nor regional administrations have unlimited resources and a free hand to pursue their own policies without reference to other administrations. We will further assume that it is axiomatic that local interests and energies should be harnessed to the greatest practicable degree, even though this does not always happen in practice.

## Regional development agencies

Most West European countries have some system of regional development agencies. Most of the accounts of their operation concentrate on the scale of resources they deploy and the range of programmes in which they engage, but ignore important aspects of the manner in which they work (Yuill (ed.), 1982). Yet there have been some significant changes in recent years, at least in Britain, which can be summed up as a two-fold trend: a growing awareness of the need to involve local communities (local government, individuals and businesses) in the planning and execution of projects; and a greater willingness to recognize the multiple facets of development work. Until the mid-1970s, Britain had three development agencies – the Mid-Wales Development Board, the Highlands and Islands Development Board (HIDB) and the Development Commission (which operated only in rural areas but had responsibilities throughout rural Great Britain). In the mid-1970s, the Welsh and Scottish Development Agencies were established, taking over the assets of the Development Commission in their respective areas. The Development Commission (recently renamed the Rural Development Commission) now operates only in rural England. The Mid-Wales Development Board and HIDB both started operation in a 'top-down' manner and maintained that mode of work well into the 1970s or even 1980s. Plans were made and implemented from their respective headquarters at Newtown and Inverness (Grassie, 1983; Wenger, 1980). In contrast, the Scottish Development Acency (SDA), virtually from its inception in the mid-1970s, became involved in locally based, comprehensive development schemes which involved many participants, although this involvement came about for reasons which included an element of the fortuitous. The Scottish Development Department had been critical of the attempts at comprehensive redevelopment in Glasgow and was looking for a more effective way of proceeding. Early in 1976, the new town project at Stonehouse was cancelled. The outcome was the establishment of the Glasgow Eastern Area Redevelopment project (GEAR), with the SDA taking the lead role and injecting significant resources, in conjunction with the local authorities and the private sector (Wannop and Leclerc, 1987). The team which had been assembled for the Stonehouse development was deployed in Glasgow. So successful was this programme perceived to be that it quickly became the model for other initiatives by the SDA. The principle is to select a particular area, to concentrate resources thereon to break the downward spiral of activity and confidence, and, after a limited number of years, having helped the locals to turn their economy around, to move elsewhere. As of 31 March 1989, there were five integrated area projects in hand, while in three other cases, including GEAR, the Agency had completed its commitment and withdrawn (SDA, 1989). By that date, also, the Agency had in place a regional structure of organization, with seven regions covering all of Scotland outside the area in which the HIDB operates.

The HIDB has made moves in the same direction, most notably in 1982 by setting up area offices, so that decisions can be taken nearer the scene of action and in conjunction with local interests. The Board also encourages multi-purpose village co-operatives and other communally based ventures; the first multi-purpose co-operative was set up in 1977.

The RDC has always taken a much more 'bottom up' approach, a fact which was made explicit in 1984. The Commission announced the designation of new rural development areas (being areas in greatest need of help) and invited the local authorities and other interests in each area to form a committee for the purpose of drawing up a programme, which would include contributions from public and private bodies as well as bids for funds from the Commission. These bids are judged against knowledge of the needs and possibilities for each area, not against a master plan or blueprint – none such has been prepared by the Commission. Other Commission expenditure, outwith the Rural Development Programme, is also largely in response to local initiatives. Since the RDP procedure was initiated, it has been copied in counties which do not have a Rural Development Area and therefore do not qualify for some of the main heads of expenditure, such as small workshops. In the next few years, the Commission will probably strengthen its regional and county presence, to facilitate its dialogue with local interests and improve its ability to respond to local needs.

During the last ten to fifteen years, the direction of change has been quite clear – towards a greater appreciation on the part of regional development agencies in Britain that successful work depends on good co-operation with local authorities and other locally based organizations. To an increasing extent, therefore, the agencies are providing reinforcement for local authorities.

## The role of local authorities

In all the democratic, industrially developed countries, local authorities are and will remain an important element in the development process, notwithstanding the diversity of patterns concerning their legal status, duties and powers. Episodically, the general pattern of local government may be changed (Chisholm, 1975); historically, the trend has been toward the enlargement of the area of jurisdictions, though their powers may be reduced or moved up the hierarchy of authorities. In particular cases, it may be necessary to abrogate the functions of local government for a period of time, as with the new towns programme in Britain (non-elective development corporations), and the application of this concept to urban situations which are deemed to require action on a scale beyond the competence of the local authority(ies) or where, as in the case of the London docklands, the relevant parties could not agree on the strategy to be pursued (Ledgerwood, 1985). In any case, the creation of this and other urban development corporations is conceived to be temporary; the eleven in existence in 1989 are all expected to have completed their tasks by the mid- to late 1990s (the two oldest, the London and Liverpool corporations, having been established in 1981). The corporations will be wound up and administration will revert to the existing or modified local authority system. More immediately, the urban development corporations generally, and the London Docklands Development Corporation in particular, have been criticized for their failure to provide some of the basic facilities which are essential if development is to be successful (Audit Commission, 1989, pp. 27–8). Indeed, the failure to plan adequate transport facilities, to provide low-cost housing and the various amenities which residents and workers require has (temporarily?) caused confidence in London's docklands to wither, leading to a collapse of property values and a crisis of confidence (*Independent*, 4 January 1990). Rather late in the day, the LDDC is reinventing a wheel which had been hastily discarded in the initial rush to develop. Other urban development corporations, as for example in Sheffield, have from the beginning behaved much more as if they were local authorities having to be concerned with the totality of local facilities.

The great majority of restructuring projects are carried out in the context of the existing local authority system. It is desirable that this should be able to cope with the ongoing changes which are inevitable; the temporary abrogation of this system should be regarded as a regrettable last resort if all else fails. Therefore, we will concentrate on the role of local authorities in facilitating the changes that are needed.

Despite the formal constitutional differences between England and West Germany, local authorities engage in a similar range of activities aimed at promoting economic development: Johnson and Cochrane (1981) list the following forms of intervention:

- provision of land and premises
- variation of land prices and rents
- loans, grants and tax reductions
- promotion and advertising
- professional advice and assistance
- mobilization of support from other levels of government
- locally determined infrastructure
- planning

The relative importance of these forms of intervention varies between the two countries and also between authorities within each nation. While the provision of land and premises is the most important activity in both countries, the German authorities are more willing to offer property sales and rentals at concession rates than are their English counterparts; in contrast, English local authorities more commonly have a direct involvement in projects. The kinds of intervention listed above are similar to those discussed by Blakely (1989) for development programmes organized by local authorities and communities within the United States.

Two other activities deserve particular mention. Local authorities usually have an important part to play in the good education of their population and development of skills of all kinds. The all-purpose local authorities (counties and boroughs) have the primary responsibility in Britain, whereas in the United States school provision is made by single-purpose school boards for their respective districts, financed by a separate tax on households. The situation in Britain is in a state of flux, with the removal in 1988 of polytechnics from local authority control, with provisions coming into force for schools to be able to opt out of local control and with the progressive establishment of training and enterprise councils to take over responsibilities for centrally funded training and retraining at the post-school level. In the second place, there has been a remarkable growth in enterprise agencies of one kind or another, whose purpose is to foster local enterprise and to generate a suitable climate.

In general terms, local initiatives in the fields identified above are consistent with the needs which modern industry has, as identified in Chapter 8. However, the real problem is one of converting these general points into practical programmes on the ground:

There are no easy solutions and no unambiguously right answers to the problems of urban regeneration and economic development: different solutions will be appropriate in different places; what works now may not work in a few months as circumstances change; a solution to one problem might simply reveal another ....

The emphasis in recent years has been on the primacy of the private sector as the engine for economic recovery. This is surely right. But its contribution is limited by shortages of land, labour and capital. The impact and extent of such shortages vary from area to area. A local strategy to identify and overcome them is needed.

To be most effective central government, through its local agencies, ought to participate in this process. To do so it needs to be better coordinated and to place less reliance on a scheme-by-scheme approval approach and more on encouraging the development of coherent local strategies.

(Audit Commission, 1989, pp. 52 and 54)

In furtherance of this aim, the Audit Commission has identified a body of good practice in strategy development and programme implementation at the local level which has been consolidated into a guide which its auditors have been trained to use with local authorities to identify the strengths and weaknesses of their economic development activities.

The first requirement is to identify the bottlenecks which inhibit the maintenance and/or expansion of employment, which is equivalent to asking what are the limitations experienced by existing firms and potential new ones (whether these are inward investors or formed locally). The second is to find ways to ease or eliminate those bottlenecks. Although there may well be general advice which can be offered, the actual pattern of needs and programmes will inevitably vary from one locality or region to another. Some examples will illustrate this proposition.

New York State used to be the premier manufacturing state of the United States but had lost that position by 1973 and thereafter, badly affected by deindustrialization, faced the possibility that, notwithstanding the growth of service employment, long-term economic decline might set in. By the close of the 1970s, it was realized that although considerable progress had been made through substantial tax reductions and incentive schemes to foster both manufacturing and service activities, nevertheless: 'the customary policy levers were no longer sufficient to revitalize an ailing economy' (Schoolman, 1986, pp. 13–14). From the early 1980s, the emphasis was switched away from macro-economic measures toward micro-economic policies designed to foster existing firms and the creation of new ones. Three such measures of particular importance were introduced to assist new firms and small businesses: substantial reduction of state taxes; assistance in the procurement of development finance; and easement of regulatory policies, which were perceived to favour large firms. In addition, special attention was paid to high technology manufacturing.

Once a state can lay claim, as New York could, to one of the most advanced university and research networks in the country, a well educated and trained labor force, an elaborately designed infrastructure, especially its system of transportation, and a functionally diverse set of financial institutions, there are few preconditions which remain to be met for a free market to prosper. Perhaps the most important remaining factor, and one originating beyond the boundaries of New York State, was ... of *organized* competition for high-technology industry.

(Schoolman, 1986, pp. 18–19; emphasis in the original)

Schoolman cites Massachusetts, North Carolina and California as the primary loci for this perceived competition in high technology.

Already having a strong base in high technology pursuits, the decision was taken to create in New York State some mechanisms which would help to improve the competitive position. In essence, the state embarked on a programme to consolidate and enhance the existing beneficial externality effects. The first step was to revitalize the somewhat moribund New York State Science and Technology Foundation; a newly funded Science and Technology foundation was created in 1981. As an umbrella organization for the stimulation of new high technology firms, the Foundation was given wide powers. Apart from direct assistance to firms, the Foundation was charged with organizing efforts to obtain for the state a larger share of federal R and D funding – the state was perceived to be getting less than its due – and was given the power to designate Centers of Advanced Technology in universities. The express purpose of these is to increase the links between manufacturing companies in high technology and academic research, links which had not evolved of their own accord to anything like the extent which was thought to be necessary. A parallel development, aimed at the wider generality of manufacturing enterprises, was the establishment, under the Urban Development Corporation, of a Center for Industrial Innovation, with the aim of improving the rate of adoption of new technology, ideas and products in the manufacturing sector, in a bid to maintain and improve the competitive position (Schoolman and Magid, 1986).

The change of attitude indicated by the measures taken, particularly in the manufacturing field, had implications also for the service sector, the most buoyant part of the state's economy. The precise impact of these measures cannot be ascertained. On the other hand, a population loss of 680,000 in the 1970–80 period became a gain of 210,000 between 1980 and 1986, and unemployment, which was above the national average, fell below it in the same six-year period. Although wages per hour in manufacturing are lower than those paid in twenty other states, New York State ranks third on average annual pay (after Arkansas and the District of Columbia) (US Bureau of the Census, 1987). At the very least, it seems probable that the measures taken have not been harmful and that in all probability they assisted in the significant turn-around in the fortunes of New York State.

Swansea, South Wales, provides a contrasting problem. The Lower Swansea Valley skirts the main urban area on the east, though it is partially surrounded by built-up areas. After the Second World War, all the steel plants closed and all but one of the zinc and copper smelters. They left behind a moonscape of tip heaps, chimneys and buildings extending over a triangular area about 1.6 kilometres along the northern base and extending about 6.4 kilometres to the southern apex (that is, about 10 sq km). The valley itself, and the surrounding slopes, were nude of any trees – they had all been killed by the noxious smelter fumes. This virtually useless tract of land was a millstone round the neck of the local economy. For that millstone to be removed, two problems, among others, had to be solved. The copper and zinc residues are harmful to most plant life and low-cost ways had to be found to cope with this problem. Second, until the complex pattern of land ownership could be established, no development proposals could be made, let alone implemented. Both problems were tackled by dedicated enthusiasts at University College Swansea (Bromley and Humphrys (eds), 1979). Once the ownership issue had been resolved and some progress made in dealing with the toxic wastes – much of which ended up being used as hardcore – some piecemeal development for manufacturing and wholesaling began in about 1960. Helped by national regional policy and then by designation in the early 1980s as an Enterprise Zone, the greater part of the formerly derelict area has now been brought back into use, to the long-term benefit of Swansea and its environs even though some of the development represents local relocations across the EZ boundary (Bromley and Morgan, 1985; Bromley and Rees, 1988).

South Wales as a whole, including Swansea, has experienced above average unemployment throughout the post-war period and, as a consequence, has benefited from the incentives available to firms under the package of regional economic policy. As of October 1989, unemployment in the Swansea travel-to-work area was 7.6 per cent compared with a national average of 5.8 per cent. Reclamation of the Lower Swansea Valley has not solved the area's problems, but it has undoubtedly helped to ameliorate them.

Two other British cities, both suffering unemployment chronically higher than the national average, illustrate somewhat different approaches. During the late 1970s and early 1980s, Sheffield experienced the decimation of its steel and steel-based industries, with resulting dereliction and abandonment in the Don Valley comparable to that in Swansea, and the associated loss of employment. An urban development corporation has been created for this area. Working with the City of Sheffield and other authorities, and taking a strategic view of the future, a bold programme is now in hand, the effect of which should be to transform the image of the area. A major new shopping centre is one element. Another is the largest complex of swimming facilities in Europe, including an olympic standard pool, plus an artificial ski slope on the flanks of the Pennines. These works are in hand or completed, while at the same time two theatres are being renovated so that they can be brought back into use. By the mid-1990s, Sheffield will have splendid sports facilities, and much improved shopping and cultural amenities. The effect will be radically to improve the attractions of the city and its environs as a place in which to live and work, which will materially assist the task of maintaining and expanding employment. Unemployment in the Sheffield travel-to-work (TTW) area in October 1989 was 8.6 per cent, serious but not disastrous.

Glasgow has had an intractable problem over much of the post-war period (11.5 per cent unemployment in October 1989 in the TTW area), with little sign of radical improvement up to the mid-1970s, notwithstanding considerable efforts. Since GEAR was initiated in 1976, there has been a notable injection of confidence; the visitor is quickly struck by the general sense that things are improving. Acquisition of the splendid Burrell collection of pictures was a major fillip, and the city has energetically sought to foster the range and quality of its cultural facilities. For the year 1990, Glasgow succeeds Paris as Europe's 'cultural capital'; some £14 million will be spent on musical and other events. In this way, Glasgow is setting out to challenge the long-held stereotype that it is the place of muck and brass, whereas Edinburgh is the Scottish city of culture and civilization. As with Sheffield, it is of course too early to say how successful this enterprise will be in fostering a new identity and creating the kind of place in which modern industry will thrive.

These cases are examples only, and incomplete examples at that. However, they serve to indicate some of the ways in which regional and local initiatives can be mobilized to improve the environment for existing and potential firms and their workers. The list of other instruments which are being employed, usually in some form of package, is long indeed: small workshop provision; science parks; joint ventures between local authorities and private interests; the establishment of enterprise trusts, often in conjunction with the local chamber of commerce; and the improvement of the local skill base; these are but a few of the measures being undertaken. There is some hope that the 'slings and arrows of outrageous fortune' can in some degree be resisted.

## Conclusion

The purpose of this chapter has been to explore the scope that exists for local initiative in solving the problems that a region may experience. In so doing, it is not intended to imply that a regional or local authority administration can 'go it alone' and, by its own actions, find all the relevant solutions. That is manifestly not true. On the other hand, it would be equally untrue to suggest that local action is entirely useless and unavailing.

Neo-classical thought places the emphasis on individual decisions taken in the light of circumstances over which they have no control. Collective action, which serves to create positive or negative externalities, is assumed away. In contrast, the Keynesian approach, dominant until the 1970s, assumed that the only really important collective action lay at the national level, in respect of the management of demand and its spatial reallocation to assist regions where demand was perceived to be 'deficient'. Although in some countries regional planning was in vogue for a time, that process never really engaged the energies and enthusiasms of local people. The importance of doing just that has become more apparent in the last one or two decades, not least as a major and essential ingredient of the supply-side approach to economic management. In some respects, however, practice at the local level has been running ahead of the willingness of central governments to recognize its importance and invent ways to foster better mechanisms for the co-operation between agencies at the national, regional and local scales which is essential for creating and maintaining a system of regions which can adapt and change without too much pain.

Local initiative, therefore, is but an ingredient in the process by which regions grow and decline. It is an ingredient which has hitherto received too little attention in the literature on regional economics.

# Unequal exchange

This doctrine is most closely associated with the work of Emmanuel (1972) and the close development thereof which Amin published just two years later. We will initially focus on Emmanuel's influential exposition. He begins his formal argument with the classic theorem used by Ricardo to determine comparative advantage in international trade, a 2×2 table of production costs for a 'unit' of output, representing two commodities in two countries. However, in presenting the exchange system in terms of the labour theory of value, Emmanuel shifts to a different presentation; Table A.1 reproduces one of his imaginary data sets, for ease of reference. The key points to note are the following. The total cost of production is shown as the same in both country A and B, at 110. However, because 'profits' differ, the total 'price' of the products differs – 190 and 150. Now it is *assumed* that B will export its production to A. If B exports one unit of output, it will receive a revenue of 150 (the price of production shown in Table A.1) with which to purchase goods from A. If we ignore all transfer costs: 'leaving aside the inputs in past labor that we assume to be the same in both countries, country B exchanges one hour of its national living labor for 15/19 hours of A's living labor' (Emmanuel, 1982, p. 161). Note that this formulation says nothing about the commodity composition of the trade, being couched in terms of some generalized commodity or basket of commodities. Note also that the terms on which trade occurs are assumed. Furthermore, it is assumed that obtaining 15/19 of an hour of A's 'national living labour' for one hour of its own confers no advantage on country B. That could only be true if the generalized commodities (commodity baskets) are identical. Since that would be widely perceived as an implausible assumption, some consideration must be given to the utility of the goods which are exchanged.

Surplus value is the product in excess of the wage for socially necessary labour. Consequently, the rate of profit (within the Marxist rules) is surplus value divided by the amount of capital invested. Now, in Table A.1, Emmanuel calculates the rate of profit for the two countries combined (120/360 = 33 1/3%). This amalgamated rate of profit is used to allocate the total profit ( = surplus value = 120) between the two countries, profit being calculated as one-third of 240 and 120 respectively. Thereby, the imputed profit is 80 for country A and 40 for country B. The assumption that profit rates must be equal in both countries derives from the assumption that capital is fully mobile, so that international and intersectoral shifts will ensure uniformity in the returns. However, the effect of this assumption is that country B receives a 'price' of 150

**Table A.1** Imaginary data to illustrate the process of unequal exchange

| Country | K<br>Total capital invested | c<br>Constant capital consumed | v<br>Variable capital | m<br>Surplus value | V<br>Value c+v+m | R<br>Cost of production c+v | T<br>Rate of profit $\frac{\Sigma m}{\Sigma K}$ | p<br>Profit T×K | L<br>Price of production R+p |
|---|---|---|---|---|---|---|---|---|---|
| A | 240 | 50 | 60 | 60 | 170 | 110 | $33\frac{1}{3}\%$ | 80 | 190 |
| B | 120 | 50 | 60 | 60 | 170 | 110 | | 40 | 150 |
| | 360 | 100 | 120 | 120 | 340 | 220 | | 120 | 340 |

*Source*: Emmanuel, 1972, p. 162

for goods with a 'value' of 170, whereas country A gains a revenue of 190 for goods valued at 170. Furthermore, to assume equality in the rates of profit is a strong assumption, every bit as strong as the Ricardian assumption that factors of production are immobile between nations.

The brief exposition serves to show two points. First, unequal exchange theorists make assumptions every bit as heroic as the neo-classicists about whom they complain. Second, in the above formulation, which can be regarded as 'mainstream' Marxist, nothing is actually said about *why* trade occurs and what the commodity composition may be. In essence it is assumed that trade is occurring, and it is further assumed that the terms of this trade are given. Amin takes the argument no further, nor, in essence, does Emmanuel, until near the end of his treatise (Chapter 6).

In this chapter, Emmanuel reverts to the classic example used by Ricardo, of cloth and wine in the two-country case of England and Portugal. Treating Ricardo's hour of work as the 'prices of production', i.e., the price in the market, he generates an imaginary set of data for variable capital, etc., which allows for the following result. Conventional specialization occurs (Portugal producing wine and England cloth), which will yield a 'price' saving for the system as a whole and also a saving in the 'hours of living labour' used: 'Everything has thus turned out for the better' (p. 247). That statement seems curious, in that, on Emmanuel's example, Portugal has had to *increase* its hours of 'living labour', by comparison with the pre-trade situation, while England has saved on hours of work. In his second example, Emmanuel allows wages in Portugal to rise by one-third, everything else remaining constant, except that the 'rate of surplus value' in Portugal drops from 100 to 50 per cent. The result is to change the 'price of production' of wine and cloth in Portugal, and thereby to reverse the pattern of comparative advantage derived from the 'price' data. In which case, if trade is to occur, Portugal makes cloth and England wine. The effect would be that the total hours of 'living labour' would increase for the system as a whole compared with the pre-trade situation, although Portugal would have gained while England lost in this respect. On the basis of this single numerical example, Emmanuel concludes:

> Consequently, if the organic composition [capital-output ratio] of the different branches is not the same ... only a certain disparity in wages ... between the different countries is needed for an international division of labor based on comparative costs to lead, not to a gain, but to a loss for the world as a whole.
>
> (Emmanuel, 1972, p. 248)

Now that is a strong conclusion to base on a single example using imaginary data. Is it a robust conclusion, a conclusion of general applicability? Emmanuel cannot have meant it to apply to a situation in which one of the commodities can be produced in only one country (e.g., localized minerals), and we will confine our discussion to the situation which he was envisaging, that both nations can produce both goods. The first point to note is that Emmanuel refers to a 'disparity in wages'. Yet, this is, in fact, based on a comparison in which the wage differences between England and Portugal were *changed*, wages in Portugal being raised by one-third. On Emmanuel's own reasoning, it is the increase in Portugal's wages to something nearer thòse of England which creates the adverse effect, since wages differed between England and Portugal in the first case. His conclusion, therefore, does not follow from his example.

The second problem is this. If, by engaging in trade on the basis of comparative advantage, a nation has to employ a larger number of hours of 'living labour', and if this truly is a loss, then why does the nation engage in trade? The comparative advantage theorem shows the pattern of *potential* gains (losses) from engaging in trade; it is not a statement that trade will actually occur. Thus, even where, as in Emmanuel's first example, the system as a whole saves hours of 'living labour', one country, Portugal, would have to expend additional hours. If, under these circumstances, trade actually occurs between free and sovereign nations, we must postulate one of two circumstances:

(a)   That there is some form of money illusion
(b)   That the analysis is incorrect.

The crux of the Marxist analysis in this context is that market prices do not reflect 'real' values with sufficient accuracy. If we equate market values with 'values in exchange', then Marxists argue that in addition there is a 'use value' and 'value' (generated by the employment of labour). The basis of all *real* 'value' is held to be the labour expended in production. Hence, the basis of the unequal exchange thesis is that there is indeed a money illusion, in that market prices do not reflect at all accurately the embodied hours of work.

There are several reasons why a theory of value based on labour expended runs into difficulties. The classical problem is that of a non-reproducible article – how is it that wine increases in value as it matures, without the expenditure of additional labour? Or how may one value a work of art? If these be regarded as exceptions which may be set to one side, what are we to say about natural resources which may be destroyed or rendered inaccessible by a proposed development? If they are virgin resources which have had no labour expended on them, do they really have no value? Most analysts would say that they do have a value, and that this should be taken into account in making a decision whether to proceed or not. Or again, because a wilderness has not experienced man's interference, does it truly have no worth? In which case, how could we justify the maintenance of areas such as the national parks of the United States?

To ask these questions is to point out one reason why the labour theory of value cannot be a wholly general theory. At best, it can be a partial theory, concerned with some (the majority?) of man's artefacts. However, within that restricted domain, there are some formidable problems, all of which revolve around the problem of measurement. First, how is one to reduce the different kinds of work – for example, the lathe operator and the design engineer – to common units of labour? How, second, are these units of labour to be related to the concept of subsistence? Even if a single subsistence wage can be visualized for one economy, it must be accepted that the subsistence level does vary from nation to nation. The concept of 'living labour' measured in hours of work looks to be highly dependent on the assumptions which are made to convert one kind of labour to another and to compare the outcome with local standards of living. The third problem concerns the use of capital equipment – 'embodied labour', as Marxists call it. To convert the market price of a machine into X hours of labour involves the use of pretty heroic assumptions.

Sheppard and Barnes (1990) claim that these problems have been solved, that: 'Marxian labour values can be measured rigorously from data that includes an input-output matrix, information on the wage bill, and estimates of market prices.' The wage bill, for example, can be converted into labour hours by dividing it by the hourly wage. Such a procedure implies that labour hours are derived values, not data of a primary nature, and that the measurement of labour hours is dependent on the use of observed market prices. In which case, labour hours cease to be an independent, objective measure of production and exchange relationships.

The doctrine of unequal exchange, reduced to its essentials, amounts to this. If two countries engage in trade, and if the productivity of A is low and that of B is high, then A may have to export goods which embody ten hours' work and receive in return goods which embody only one hour's work in B. It could be said that A has a net loss of 9 hours, balanced by B's gain of the equivalent number. On that basis, exchange is held to be unequal. However, given the productivity differences and associated differences in wages, the exchange ratio reflects nothing but the characteristics of the two economies. And if the use value of the articles as exchanged is sufficient, the trade partners would indeed be better off with trade than without.

Since the appearance of the Amin and Emmanuel volumes in the early 1970s, numerous scholars have continued to wrestle with the doctrine of unequal exchange. However, as Sheppard and Barnes (1990) note, the theoretical formulation of the doctrine is open to criticism on several fronts, and they end their discussion by shifting the argument into a different form:

> if the money wage in a region increases relative to wage levels in other regions, and if the current geographical patterns of specialization and trade remain profitable, then that region will increase its monetary gains from trade. This is because the production prices of its export will increase relative to the production prices of imports.
>
> (Sheppard and Barnes, 1990)

The implicit assumption is that price changes have no impact on the quantity of production and trade; this is not realistic. Therefore, although the above statement is logically correct, it is based on a false premiss.

Edwards (1985) recognizes the problems which have been noted above and interprets the unequal exchange thesis in terms that shift the argument considerably. Instead of postulating a net transfer of resources from the poor country to the rich one, he argues that the real complaint is about an unequal sharing of the benefits to be had from trade:

The exchange is unequal because the poor countries exchange goods in which more labour-time is embodied for goods in which less labour-time is embodied. The actual situation is compared with a *hypothetical* one in which wages are equal in the rich and poor countries.... Emmanuel's theory is ... a complaint about the *distribution* of gains from international trade.

(Edwards, 1985, pp. 63 and 68)

For all the apparent solidity of the unequal exchange thesis in providing a mechanism for the systematic transfer of wealth from the poor to rich nations, closer inspection shows it to be based on the assumption that an hour's work has the same worth irrespective of the circumstances under which that work is done. If that assumption is not accepted, the thesis fails. In contrast, if unequal exchange is interpreted in the manner suggested by Edwards, then it does not differ from conventional trade theory, which recognizes limiting terms of trade but does not specify where within that range actual exchange will occur. If trade is between atomistic producers and consumers behaving competitively, mutual benefit should occur. If, as is more realistically the case, trade is influenced by state intervention on the one hand, and the policies of multi-national companies on the other, the terms on which trade is conducted will depend on the relative bargaining power of the respective trade partners. There is, therefore, no 'necessity' by which the balance of advantage must always be biased in one direction, and the nature of any such bias becomes a matter for empirical investigation rather than for theoretically based assertion.

# References

Agnew, J. (1987) *The United States in the world-economy. A regional geography*, Cambridge University Press, Cambridge.

Albrechts, L., F. Moulaert, P. Roberts and E. Swyngedouw (eds) (1989) *Regional policy at the crossroads: European perspectives*, Jessica Kingsley, London.

Aldcroft, D. H. (1984) *Full employment: the elusive goal*, Wheatsheaf Books, Brighton.

Allen, G. C. (1929) *The industrial development of Birmingham and the Black Country, 1860–1927*, Allen & Unwin, London.

Allen, G. C. (1959) *British industries and their organization*, Longman, London.

Allen, K. J. (1969) 'The regional multiplier: some problems of estimation', in J. B. Cullingworth and S. C. Orr (eds) *Regional and urban studies. A social science approach*, Allen & Unwin, London, 80–96.

Allen, K. J., J. Bachtler and D. Yuill (1988) 'Alternatives in regional incentive policy design', *Regional and Industrial Policy Research Series*, No. 1, University of Strathclyde, Glasgow.

Alonso, W. (1978) 'A theory of movements', in N. M. Hansen (ed.) *Human settlement systems. International perspectives on structure change and public policy*, Ballinger, Cambridge, 197–211.

Amin, A. and J. Goddard (eds) (1986) *Technological change, industrial restructuring and regional development*, Allen & Unwin, London.

Amin, J. (1974) *Accumulation on a world scale. A critique of the theory of underdevelopment*, vol. 1, Monthly Review Press, New York.

Anderson, J., S. Duncan and R. Hudson (eds) (1983) *Redundant spaces in cities and regions? Studies in industrial decline and social change*, Academic Press, London.

Anderson, W. P. and D. L. Rigby (1989) 'Estimating capital stocks and capital ages in Canada's regions', *Regional Studies*, 23, 117–26.

Andrews, R. B. (1953–6) 'Mechanics of the urban economic base', twelve papers in *Land economics*, vols 29 to 32.

Archer, B. H. (1976) 'The anatomy of a multiplier', *Regional Studies*, 10, 71–7.

Archibald, G. C. (1967) 'Regional multiplier effects in the UK', *Oxford Economic Papers*, 19, 22–45.

Armstrong, H. and J. Taylor (1978) *Regional economic policy and its analysis*, Philip Allan, Oxford.

Armstrong, H. and J. Taylor (1985) *Regional economics and policy*, Philip Allan, Oxford.

Arndt, S. W. and L. Bouton (1987) *Competitiveness. The United States in world trade*, American Enterprise Institute for Public Policy Research, Washington.

Ashcroft, B. (1982) 'The measurement of the impact of regional policies in Europe: a survey and critique', *Regional Studies*, 16, 287–305.

Audit Commission for Local Authorities in England and Wales (1989) *Urban regeneration and economic development. The local government dimension*, HMSO, London.

Aydalot, P. and D. Keeble (eds) (1988) *High technology industry and innovative environments: the European experience*, Routledge, London.

Baker, J. N. L. and E. W. Gilbert (1944) 'The doctrine of an axial belt of industry in England', *Geographical Journal*, CIII, 49–63.

Balassa, B. (1966) 'Tariff reductions and trade in manufactures among the industrial countries', *American Economic Review*, 56, 466–73.

Barber, J. and G. White (1987) 'Current policy practice and problems from a UK perspective', in P. Dasgupta and P. Stoneman (eds) *Economic policy and technological performance*, Cambridge University Press, Cambridge, 24–50.

Becker, G. S. (1964) *Human capital. A theoretical and empirical analysis, with special reference to education*, National Bureau of Economic Research, New York.

Bentham, G. (1985) 'Trends in the relationship between earnings and unemployment in the counties of Great Britain, 1978 to 1983', *Area*, 17, 267–75.

Berry, B. J. L. (1972) 'Hierarchical diffusion: the basis of developmental filtering and spread in a system of growth centers', in N. M. Hansen (ed.) *Growth centers in regional economic development*, The Free Press, New York, 108–38.

Blackaby, F. (ed.) (1981) *De-industrialisation*, Heinemann, London. First published 1978.

Blakely, E. J. (1989) *Planning local economic development. Theory and practice*, Sage, Newbury Park, Calif.

Bluestone, B. and B. Harrison (1982) *The deindustrialization of America: plant closings, community abandonment, and the dismantling of basic industry*, Basic Books, New York.

Bolton, J. E. (1971) *Report of the Committee of Enquiry on Small Firms*, HMSO, London.

Borts, G. H. and J. L. Stein (1964) *Economic growth in a free market*, Columbia University Press, New York.

Breheny, M. J. and P. Congdon (eds) (1989) *Growth and change in a core region: the case of southeast England*, Pion, London.

Breheny, M. J. and R. McQuaid (eds) (1987) *The development of high technology industries: an international survey*, Croom Helm, London.

Brittan, S. (1983) *The role and limits of government. Essays in political economy*, Temple Smith, London.

Britton, J. N. H. (1989) 'Innovation policies for small firms', *Regional Studies*, 23, 167–73.

Bromley, R. D. F. and G. Humphrys (eds) (1979) *Dealing with dereliction. The redevelopment of the Lower Swansea Valley*, University College of Swansea, Swansea.

Bromley, R. D. F. and R. H. Morgan (1985) 'The effects of enterprise zone policy: evidence from Swansea', *Regional Studies*, 19, 403–13.

Bromley, R. D. F. and J. C. M. Rees (1988) 'The first five years of the Swansea Enterprise Zone: an assessment of change', *Regional Studies*, 22, 263–75.

Brotchie, J. F., P. Hall and P. W. Newton (eds) (1987) *The spatial impact of technological change*, Croom Helm, London.

Browett, J. and R. Leaver (1989) 'Shifts in the global capitalist economy and the national economic domain', *Australian Geographical Studies*, 27, 31–46.

Brown, A. J. (1972) *The framework of regional economics in the United Kingdom*, Cambridge University Press, Cambridge.

Brown, A. J. and E. M. Burrows (1977) *Regional economic problems. Comparative experiences of some market economies*, Allen & Unwin, London.

Buswell, R. J., R. P. Easterbrook and C. S. Morphet (1985) 'Geography, regions and research and development activity: the case of the United Kingdom', in A. T. Thwaites and R. P. Oakey (eds) *The regional economic impact of technological change*, Frances Pinter, London, 36–66.

Caesar, A. A. L. (1964) 'Planning and the geography of Great Britain', *Advancement of Science*, XXI, 1–11.

Cairncross, A. (ed.) (1971) *Britain's economic prospects reconsidered*, Allen & Unwin, London.

Casetti, E. (1981) 'A catastrophe model of regional dynamics', *Annals*, Association of American Geographers, 71, 572–9.

Castells, M. (ed.) (1985) *Urban affairs annual reviews, 28: high technology, space and society*, Sage, Beverly Hills.

Central Statistical Office (1989) *Financial statistics*, 331, November, HMSO, London.

Chamberlin, E. H. (1933) *The theory of monopolistic competition. A reorientation of the theory of value*, Harvard University Press, Cambridge, Mass. (Eighth edition published in 1962).

Chandler, J. A. and P. Lawless (1985) *Local authorities and the creation of employment*, Gower, Aldershot.

Chapman, K. and G. Humphrys (eds) (1987) *Technical change and industrial policy*, Blackwell, Oxford.

Cheshire, P. C. (1990) 'Causal factors in European urban change 1971 to 1988', *Discussion Papers in Urban and Regional Economics*, 44, University of Reading.

Chinitz, B. (1961) 'Contrasts in agglomeration: New York and Pittsburgh', *American Economic Review. Papers and Proceedings*, 51, 279–89.

Chinitz, B. (1966) 'Appropriate goals for regional economic policy', *Urban Studies*, 3, 1–7.

Chisholm, M. (1964) 'Must we all live in southeast England?', *Geography*, XLIX, 1–14.

Chisholm, M. (1966) *Geography and economics*, Bell, London.

Chisholm, M. (1975) 'The reformation of local government in England', in R. Peel, M. Chisholm and P. Haggett (eds) *Processes in physical and human geography. Bristol essays*, 305–18, Heinemann, London.

Chisholm, M. (1976) 'Regional policies in an era of slow population growth and higher unemployment', *Regional Studies*, 10, 201–13.

Chisholm, M. (1982) *Modern world development. A geographical perspective*, Hutchinson, London.

Chisholm, M. (1985a) 'De-industrialization and British regional policy', *Regional Studies*, 19, 301–13.

Chisholm, M. (1985b) 'Better value for money? Britain's 1984 regional industrial policy package', *Environment and Planning* C, 3, 111–19.

Chisholm, M. (1987) 'Regional development: the Reagan–Thatcher legacy', *Environment and Planning* C, 5, 197–218.

Chisholm, M. (1990) 'The increasing separation of production and consumption', in B. L. Turner, W. C. Clark, R. W. Kates, J. F. Richards, J. T. Mathews and W. B. Meyer (eds) *The Earth as transformed by human action*, Cambridge University Press, Cambridge.

Christaller, W. (1966) *Central places in southern Germany*, Prentice-Hall, Englewood Cliffs. (First published in German 1933).

Christy, C. V. and R. G. Ironside (1987) 'Promoting "High Technology" industry: location factors and public policy', in K. Chapman and G. Humphrys (eds) *Technical change and industrial policy*, Blackwell, Oxford, 233–52.

Clark, C. (1940) *The conditions of economic progress*, Macmillan, London.

Clark, C. (1966) 'Industrial location and economic potential', *Lloyds Bank Review*, October, 1–17.

Clark, C. (1967) *Population growth and land use*, Macmillan, London.

Clark, C., F. Wilson and J. Bradley (1969) 'Industrial location and economic potential in Western Europe', *Regional Studies*, 3, 197–212.

Clark, G. L., M. S. Gertler and J. Whiteman (1986) *Regional dynamics. Studies in adjustment theory*, Allen & Unwin, Boston.

Cleary, M. N. and G. D. Hobbs (1984) 'The fifty year cycle: A look at the empirical evidence', in C. Freeman (ed.) *Long waves in the world economy*, Frances Pinter, London, 164–82.

Coates, D. and J. Hillard (eds) (1986) *The economic decline of modern Britain*, Wheatsheaf, Brighton.

Coddington, A. (1983) *Keynesian economics. The search for first principles*, Allen & Unwin, London.

Cole, K., J. Cameron and C. Edwards (1983) *Why economists disagree: the political economy of economics*, Longman, London.

Commission of the European Communities (1984) *The regions of Europe. Second periodic report*, Office for Official Publications of the European Communities, Luxembourg.

Commission of the European Communities (1987) *Third periodic report from the Commission on the social and economic situation and development of the regions of the Community*, Commission of the European Communities, Brussels.

Confederation of British Industry (1989) *Trade routes to the future. Meeting the transport infrastructure needs of the 1990s*, CBI, London.

Congdon, P. (1989) 'Modelling migration flows between areas: an analysis for London using the census and OPCS longitudinal study', *Regional Studies*, 23, 87–103.

Corbridge, S. (1986) *Capitalist world development: a critique of radical development geography*, Macmillan, London.

Crang, P. and R. L. Martin (forthcoming) 'Mrs Thatcher's vision of the "new Britain" and the other sides of the "Cambridge phenomenon"', *Environment and Planning*, D.

Creigh, S. W. (1979) 'Work stoppages in the Humberside sub-region: a case study in regional strike-proneness', *Regional Studies*, 13, 73–89.

Cross, M. (1981) *New firm formation and regional development*, Gower, Farnborough.

Curran, J. and J. Stanworth (1982) 'Bolton ten years on – a research inventory and critical review', in J. Stanworth, A. Westrip, D. Watkins and J. Lewis (eds) *Perspectives on a decade of small business research. Bolton ten years on*, Gower, Aldershot, 3–27.

Currie, J. (1985) *Science parks in Britain: their role for the late 1980s*, CSP Economic Publications, Cardiff.

Damette, F. (1980) 'The regional framework of monopoly exploitation: new problems and trends', in J. Carney, R. Hudson and J. Lewis (eds) *Regions in crisis. New perspectives in European regional theory*, Croom Helm, London, 76–92.

Daniels, P. W. (1985) *Service industries: a geographical appraisal*, Methuen, London.

Danson, M. (ed.) (1986) *Redundancy and recession. Restructuring the regions?*, Geo Books, Norwich.

Davies, S. (1979) *The diffusion of process innovations*, Cambridge University Press, Cambridge.

Deane, P. (1978) *The evolution of economic ideas*, Cambridge University Press, Cambridge.

Denison, E. F. (1967) *Why growth rates differ: postwar experience in nine western countries*, Brookings Institution, Washington.

Dennis, R. (1978) 'The decline of manufacturing employment in Greater London: 1966–74', *Urban Studies*, 15, 63–73.

Department of Trade and Industry (1985) *Burdens on business. Report of a scrutiny of administrative and legislative requirements*, HMSO, London.

Díaz-Alejandro, C. F. (1978) 'Delinking north and south: Unshackled or unhinged?', in A. Fishlow, C. F. Díaz-Alejandro, R. R. Fagen and R. D. Hansen (eds) *Rich and poor nations in the world economy*, McGraw-Hill, New York, 87–162.

Dickinson, R. E. (1947) *City region and regionalism: a geographical contribution to human ecology*, Kegan Paul, London.

Dixon R. and A. P. Thirlwall (1975) 'A model of regional growth rate differences on Kaldorean lines', *Oxford Economic Papers*, 27, 201–14.

Donnison, D. and M. Middleton (eds) (1987) *Regenerating the inner city: Glasgow's experience*, Routledge & Kegan Paul, London.

Dow, J. C. R. (1964) *Management of the British economy, 1945–60*, Cambridge University Press, Cambridge.

Dow, J. C. R. and I. D. Saville (1988) *A critique of monetary policy: theory and British experience*, Clarendon Press, Oxford.

Dow, S. C. (1985) *Macroeconomic thought. A methodological approach*, Blackwell, Oxford.

Drucker, P. F. (1981) 'Toward the next economics', in D. Bell and I. Kristol (eds) *The crisis in economic theory*, Basic Books, New York, 4–18.

Duijn, van J. J. (1983) *The long wave in economic life*, Allen & Unwin, London.

Dunford, M. (1988) *Capital, the state, and regional development*, Pion, London.

Dunnett, P. J. S. (1980) *The decline of the British motor industry. The effects of government policy, 1945–79*, Croom Helm, London.

Dunning, J. H. (1981) *International production and the multinational enterprise*, Allen & Unwin, London.

Edwards, C. (1985) *The fragmented world: competing perspectives on trade, money and crisis*, Methuen, London.

Elbaum, B. and W. Lazonick (eds) (1986) *The decline of the British economy*, Clarendon Press, Oxford.

Emmanuel, A. (1972) *Unequal exchange. A study of the imperialism of trade*, Monthly Review Press, New York.

Engle, R. F. (1974) 'A disequilibrium model of regional investment', *Journal of Regional Science*, 14, 367–76.

Engle, R. F. (1979) 'The regional response to factor supplies: estimates for the Boston SMSA', in W. C. Wheaton (ed), *Interregional movements and regional growth*, The Urban Institute, Washington, DC., 157–96.

Estall, R. C. (1985) 'Stock control in manufacturing: the just-in-time system and its locational implications', *Area*, 17, 129–33.

Feinstein, C. H. (1983) 'Introduction', in C. H. Feinstein (ed.) *The managed economy, Essays in British economic policy and performance since 1929*, Oxford University Press, Oxford, 1–30.

Feldman, M. M. A. (1986) 'Firm size and local employment change: three issues', *Regional Studies*, 20, 73–89.

Feulner, E. J. (1983) 'Epilogue', in M. L. Greenhut and C. T. Stewart (eds) *From basic economies to supply side economics*, University Press of America , Lanham, 263–4.

Firn, J. R. and D. Roberts (1984) 'High technology industries', in N. Hood and S. Young (eds) *Industry, policy and the Scottish economy*, Edinburgh University Press, Edinburgh, 288–325.

Fisher, A. G. B. (1935) *The clash of progress and security*, Macmillan, London.

Florence, P. S. (1948) *Investment, location and size of plant*, National Institute of Economic and Social Research, London (published by Cambridge University Press, Cambridge).

Florence, P. S. (1962) *Post-war investment, location and size of plant*, Occasional Paper XIX, National Institute of Economic and Social Research, London (published by Cambridge University Press, Cambridge).

Flynn, N. and A. P. Taylor (1986) 'Inside the rust belt: an analysis of the West Midlands economy', *Environment and Planning*, A, 18, 865–900 and 999–1028.

Foreman-Peck, J. S. (1985) 'Seedcorn or chaff? New firm formation and the performance of the interwar economy', *Economic History Review*, XXXVIII, 402–22.

Foster, J. and J. Mulley (1988) *Assessing the impact of foreign direct investment on the Scottish economy: some regional macroeconomic evidence*, Centre for Urban and Regional Research, Discussion Paper 33, Glasgow.

Fothergill, S. and G. Gudgin (1979) 'Regional employment change: a sub-regional explanation', *Progress in Planning*, 12, 155–219.

Fothergill, S. and G. Gudgin (1982) *Unequal growth. Urban and regional employment change in the UK*, Heinemann, London.

Fox, K. (1986) *Metropolitan America: urban life and urban policy in the United States 1940–1980*, University Press of Mississippi, Jackson.

Frank, A. G. (1964) 'On the mechanisms of imperialism: the case of Brazil', *Monthly Review*, 16, 284–97.

Frank, A. G. (1966) 'The development of underdevelopment', *Monthly Review*, 18, 17–31.

Frank, A. G. (1967) *Capitalism and underdevelopment in Latin America. Historical studies of Chile and Brazil*, Monthly Review Press, New York.

Freeman, C. (ed.) (1983) *Long waves in the world economy*, Butterworth, London.

Friedman, D. (1983) 'Beyond the age of Ford: the strategic basis of the Japanese success in automobiles', in J. Zysman and L. Tyson (eds) *American industry in international competition. Government policies and corporate strategies*, Cornell University Press, Ithaca, 350–90.

Friedman, M. (1956) 'The quantity theory of money – a restatement', in M. Friedman (ed.) *Studies in the quantity theory of money*, University of Chicago Press, Chicago, 3–21.

Friedman, M. (1985) 'Monetarism in rhetoric and in practice', in A. Ando, H. Eguchi, R. Farmer and Y. Suzuki (eds) *Monetary policy in our time*, MIT Press, Cambridge, Mass., 15–28.

Friedmann, J. and C. Weaver (1979) *Territory and function. The evolution of regional planning*, Arnold, London.

Frobel, F., J. Heinrichs, and D. Kreye (1980) *The new international division of labour*, Cambridge University Press, Cambridge.

Gaile, G. L. (1980) 'The spread-backwash concept', *Regional Studies*, 14, 15–25.

Gamble, A. (1981) *Britain in decline*, Macmillan, London.

Gardner, N. (1987) *Decade of discontent. The changing British economy since 1973*, Blackwell, Oxford.

Gatti, J. F. (ed.) (1981) *The limits to government regulation*, Academic Press, New York.

General Agreement on Tariffs and Trade (GATT) (1986) *International Trade 85–86*, GATT, Geneva.

Gertler, M. S. (1986) 'Discontinuities in regional development', *Environment and Planning*, D, 71–84.

Gertler, M. S. (1988) 'The limits to flexibility: comments on the post-Fordist vision of production and its geography', *Transactions*, Institute of British Geographers, 13, 419–32.

Gibbs, D. C. (ed.) (1989) *Government policy and industrial change*, Routledge, London.

Gibbs, D. C. and A. Edwards (1983) 'Some preliminary evidence for the interregional diffusion of selected process innovations', in A. Gillespie (ed.) *Technological change and regional development*, Pion, London, 23–35.

Gibbs, D. C. and A. Edwards (1985) 'The diffusion of new production innovations in

British industry', in A. T. Thwaites and R. P. Oakey (eds) *The regional economic impact of technological change*, Frances Pinter, London, 132–63.

Gilder, G. (1981) *Wealth and poverty*, Buchan & Enright, London.

Gillespie, A. (ed.) (1983) *Technological change and regional development*, Pion (London Papers in Regional Science), London.

Gilpin, A. (1973) *Dictionary of economic terms*, Butterworth, London.

Glasmeier, A. (1988) 'Factors governing the development of high tech industry agglomerations; a tale of three cities', *Regional Studies*, 22, 287–301.

Glickman, N. J. (1982) 'Using empirical models for regional policy analysis', in M. Albegov, A. E. Andersson and F. Snickars (eds) *Regional development modeling: theory and practise*, North-Holland, Amsterdam, 85–104.

Goddard, J., A. Thwaites and D. Gibbs (1986) 'The regional dimension to technological change in Great Britain', in A. Amin and J. Goddard (eds) *Technological change, industrial restructuring and regional development*, Allen & Unwin, London, 140–56.

Goldstein, J. S. (1988) *Long cycles. Prosperity and war in the modern age*, Yale University Press, New Haven.

Gordon, I. R. (1973) 'The return of regional multipliers: a comment', *Regional Studies*, 7, 257–62.

Gottmann, J. (1961) *Megalopolis: the urbanized northeastern seaboard of the United States*, Twentieth Century Fund, New York.

Gould, A. and D. Keeble (1984) 'New firms and rural industrialization in East Anglia', *Regional Studies*, 18, 189–201.

Graham, J., K. Gibson, R. Horath and D. M. Shakow (1988) 'Restructuring in US manufacturing: the decline of monopoly capitalism', *Annals*, Association of American Geographers, 78, 473–90.

Grassie, J. (1983) *Highland experiment: the story of the Highlands and Islands Development Board*, Aberdeen University Press, Aberdeen.

Green, A. E. and D. W. Owen (1989) 'The changing geography of occupations in engineering in Britain, 1978–1987', *Regional Studies*, 23, 27–42.

Green, F., (ed.) (1989) *The restructuring of the UK economy*, Harvester Wheatsheaf, Hemel Hempstead.

Greenaway, D. (1983) *International trade policy. From tariffs to the new protectionism*, Macmillan, London.

Greenhut, M. L. (1956) *Plant location in theory and in practise. The economics of space*, University of North Carolina Press, Chapel Hill.

Greenhut, M. L. (1963) *Microeconomics and the space economy. The effectivness of an oligopolistic market economy*, Scott Foreman, Chicago.

Greenhut, M. L. (1970) *A theory of the firm in economic space*, Appleton-Century-Crofts, New York.

Greenhut, M. L., G. Norman and C.-S. Hung (1987) *The economics of imperfect competition. A spatial approach*, Cambridge University Press, Cambridge.

Greenhut, M. L. and H. Ohta (1975) *Theory of spatial pricing and market areas*, Duke University Press, Durham, North Carolina.

Greenwood, M. (1975) 'Research on international migration in the United States: a survey', *Journal of Economic Literature*, 13, 397–433.

Grimwade, N. (1989) *International trade. New patterns of production and investment*, Routledge, London.

Gripaios, P. and C. Herbert (1987) 'The role of new firms in economic growth: some evidence from south west England', *Regional Studies*, 21, 270–3.

Gudgin, G. and S. Fothergill (1984) 'Geographical variation in the rate of formation of new manufacturing firms', *Regional Studies*, 18, 203–6.

Gudgin, G., M. Hart, J. Fagg, E. D.'Arcy and R. Keegan (1989) *Job generation in manufacturing industry 1973–86. A comparison of Northern Ireland with the Republic of Ireland and the English Midlands*, Northern Ireland Economic Research Centre, Belfast.

Hägerstrand, T. (1967) *Innovation diffusion as a spatial process*, University of Chicago Press, Chicago. (first published in Swedish 1953).

Hagey, M. J. and E. J. Malecki (1986) 'Linkages in high-technology industries: a Florida case study', *Environment and Planning* A, 18, 1477–98.

Haggett, P. (1965) *Locational analysis in human geography*, Arnold, London.

Hague, D. C. and P. K. Newman (1952) *Costs in alternative locations: the clothing industry*, National Institute of Economic and Social Research, London, Occasional Paper XV (published by Cambridge University Press, Cambridge).

Hall, P. (1962) *The industries of London since 1861*, Hutchinson, London.

Hall, P. (1981) 'The geography of the fifth Kondratieff', *New Society*, March 26, 535–7.

Hall, P. (1985) 'The geography of the fifth Kondratieff', in P. Hall and A. Markusen (eds) *Silicon landscapes*, Allen & Unwin, Boston, 1–19.

Hall, P. (1987a) 'The anatomy of job creation: nations, regions and cities in the 1960s and 1970s', *Regional Studies*, 21, 95–106.

Hall, P. (1987b) 'The geography of high technology: an Anglo-American comparison', in J. F. Brotchie, P. Hall and P. W. Newton (eds) *The spatial impact of technological change*, Croom Helm, London, 141–56.

Hall, P., M. Breheny, R. McQuaid and D. Hart (1987) *Western sunrise. The genesis and growth of Britain's major high tech corridor*, Allen & Unwin, London.

Hall, P. and A. Markusen (eds) (1985) *Silicon landscapes*, Allen & Unwin, Boston.

Hamilton, F. E. I. (ed.) (1974) *Spatial perspectives on industrial organization and decision-making*, Wiley, London.

Hansen, N. M. (ed.) (1972) *Growth centers in regional economic development*, The Free Press, New York.

Hansen, N. M. (1982) 'Large-city decline and innovation diffusion in the urban system of the United States', in W. Buhr and P. Friedrich (eds) *Planning under stagnation*, Nomos verlagsgesellschaft, Baden-Baden, 207–20.

Harcourt, G. C. (1972) *Some Cambridge controversies in the theory of capital*, Cambridge University Press, Cambridge.

Harris, C. (1954) 'The market as a factor in the localization of industry in the United States', *Annals*, Association of American Geographers, XLIV, 315–48.

Harrison, B. (1984) 'Regional restructuring and "good business climates": the economic transformation of New England since World War II', in L. Sawers and W. K. Tabb (eds) *Sunbelt/snowbelt. Urban development and regional restructuring*, Oxford University Press, New York, 48–96.

Harvey, D. (1982) *The limits to capital*, Blackwell, Oxford.

Hedlund, S. (ed.) (1987) *Incentives and economic systems*, New York University Press, New York.

Hepple, L. W. (1989) 'Destroying local leviathans and designing landscapes of liberty? Public choice theory and the poll tax', *Transactions*, Institute of British Geographers, 14, 387–99.

Hicks, D. A. (1986) 'Industrial renewal through technology upgrading in the US metalworking industry', in J. Rees (ed.) *Technology, regions and policy*, Rowman & Littlefield, Totowa, 218–48.

Higgins, B. J. and D. J. Savoie (eds) (1988) *Regional economic development: essays in honour of François Perroux*, Unwin Hyman, London.

Hirschman, A. O. (1958) *The strategy of economic development*, Yale University Press, New Haven.

Holland, S. (1976a) *Capital versus the regions*, Macmillan, London.

Holland, S. (1976b) *The regional problem*, Macmillan, London.

Hood, N. and S. Young (1982) *Multinationals in retreat. The Scottish experience*, Edinburgh University Press, Edinburgh.

Hoover, E. M. (1948) *The location of economic activity*, McGraw-Hill, New York.

Hotelling, H. (1929) 'Stability in competition', *Economic Journal*, XXXIX, 41–57.

Howells, J. R. L. (1984) 'The location of research and development: some observations and evidence from Britain', *Regional Studies*, 18, 13–29.

Howells, J. R. and D. R. Charles (1989) 'Research and technological development and regional policy: a European perspective', in D. C. Gibbs (ed.) *Government policy and industrial change*, Routledge, London, 23–54.

Hudson, R. (1986) 'Producing an industrial wasteland: capital, labour and the state in north-east England', in R. Martin and B. Rowthorn (eds) *The geography of de-industrialisation*, Macmillan, Basingstoke, 169–213.

Hudson, R. (1989) *Wrecking a region: state policies, party politics and regional change in North East England*, Pion, London.

Hufbauer, G. C. (1966) *Synthetic materials and the theory of international trade*, Duckworth, London.

Hughes, G. and B. McCormick (1981) 'Do council housing policies reduce migration between regions?', *Economic Journal*, 91, 919–37.

Hulten, C. R. and R. M. Schwab (1984) 'Regional productivity growth in US manufacturing: 1951–78', *American Economic Review, Papers and Proceedings*, 74, 152–62.

Innis, H. A. (1930) *The fur trade in Canada. An introduction to Canadian economic history*, Yale University Press, New Haven.

Innis, H. A. (1936) *Settlement and the mining frontier*, vol. 9, Pt. 2 of W. A. Mackintosh and W. L. G. Joerg (eds) *Canadian frontiers of settlement*, Macmillan, Toronto.

Isard, W. (1951) 'Interregional and regional input-output analysis: a model of space-economy', *Review of Economics and Statistics*, 33, 318–28.

Isard, W. (1956) *Location and space-economy. A general theory relating to industrial location, market areas, land use, trade and urban structure*, MIT Press, Cambridge, Mass.

Isard, W., D. F. Bramhall, G. A. P. Carrothers, J. H. Cumberland, L. N. Moses, D. O. Price and E. W. Schoole (1966) *Methods of regional analysis: an introduction to regional science*, MIT Press, Cambridge, Mass.

Jewkes, J. (1960) 'Are the economies of scale unlimited?', in E. A. G. Robinson (ed.) *The economic consequences of the size of nations*, Macmillan, London, 95–116.

Johnes, G. and T. J. Hyclak (1989) 'Wage inflation and unemployment in Europe: the regional dimension', *Regional Studies*, 23, 19–26.

Johns, R. A. (1985) *International trade theories and the evolving international economy*, St Martin's Press, New York.

Johnson, N. and A. Cochrane (1981) *Economic policy-making by local authorities in Britain and West Germany*, Allen & Unwin, London.

Jong, de M. W. (1987) 'New economic activities and regional dynamics', *Nederlandse Geografische Studies*, 38, University of Amsterdam, Amsterdam.

Jorgenson, D. W., F. M. Gollop and B. M. Fraumeni (1987) *Productivity and US economic growth*, Harvard University Press, Cambridge, Mass.

Joseph, R. A. (1989) 'Technology parks and their contribution to the development of technology-oriented complexes in Australia', *Environment and Planning*, C, 7, 173–92.

Kaldor, N. (1970) 'The case for regional policies', *Scottish Journal of Political Economy*, 17, 337–48.

Kaldor, N. (1971) 'Conflicts in national economic objectives', *Economic Journal*, 81, 1–16.

Kaldor, N. (1972) 'The irrelevance of equilibrium economics', *Economic Journal*, 82, 1237–55.

Kaldor, N. (1975) 'What is wrong with economic theory', *Quarterly Journal of Economics*, 89, 347–57.

Kaldor, N. (1981) 'Comment' [on Sir Alec Cairncross] in F. Blackaby (ed.) *De-industrialisation*, Heinemann, London, 18–25.

Karaska, G. J. (1969) 'Manufacturing linkages in the Philadelphia economy: some evidence of external agglomeration forces', *Geographical Analysis*, 1, 354–69.

Kaufman, B. E. (1988) 'The postwar view of labor markets and wage determination', in B. E. Kaufman (ed.) *How labor markets work*, Lexington Books, Lexington, 145–203.

Keating, M. and R. Boyle (1986) *Remaking urban Scotland. Strategies for local economic development*, Edinburgh University Press, Edinburgh.

Keeble, D. E. (1969) 'Local industrial linkage and manufacturing growth in outer London', *Town Planning Review*, 40, 163–88.

Keeble, D. E. (1986) 'Introduction', in D. E. Keeble and E. Wever (eds) *New firms and regional development in Europe*, Croom Helm, London, 1–34.

Keeble, D. E. (1988) 'High-technology industry and local environments in the United Kingdom', in I. Aydalot and D. E. Keeble (eds) *High technology industry and innovative environments: the European experience*, Routledge, London, 65–98.

Keeble, D. E. (1989) 'High-technology industry and regional development in Britain: the case of the Cambridge phenomenon', *Environment and Planning*, C, 7, 153–72.

Keeble, D. E. (1990a) 'De-industrialization, new industrialization processes and regional restructuring in the European Community', in P. Jones and T. Wild (eds) *De-industrialization and new industrialization in Britain and West Germany*, Anglo-German Foundation, London.

Keeble, D. E. (1990b) 'New firms and regional economic development: experience and impacts in the 1980s', *Cambridge Regional Review*, 1.

Keeble, D. E., P. L. Owens and C. Thompson (1982) 'Regional accessibility and economic potential in the European Community', *Regional Studies*, 16, 419–32.

Keeble, D. E. and E. Wever (eds) (1986) *New firms and regional development in Europe*, Croom Helm, London.

Kelly, T. (1987) *The British computer industry. Crisis and development*, Croom Helm, London.

Keynes, J. M. (1936) *The general theory of employment, interest and money*, Macmillan, London.

Keynes, J. M. (1937) 'How to avoid a slump', *Times*, 12–14 January. Reprinted in D. Moggridge (ed.) *The collected writings of John Maynard Keynes*, vol. XXI, Cambridge University Press, Cambridge, 384–95.

Kleinknecht, A. (1987) *Innovation patterns in crisis and prosperity*, St Martin's Press, New York.

Knapp, B. van der and E. Wever (eds) (1987) *New technology and regional development*, Croom Helm, London.

Knight, F. H. (1921) *Risk, uncertainty and profit*, Houghton Mifflin, Boston, Mass.

Kondratieff, N. D. (1935) 'The long waves in economic life', *Review of Economics and Statistics*, 17:6, 105–15.

Korte, W. B. (1986) 'Small and medium-sized establishments in Western Europe', in D. E. Keeble and E. Wever (eds) *New firms and regional development in Europe*, Croom Helm, London, 35–53.

Krebs, G. and R. Bennett (1989) *Local economic development initiatives in Germany*, Research Papers, Department of Geography, London School of Economics, London.

Kronsjö, T. (1987) 'British work incentives and the IR/DHSS effective tax system: an essay designed to provoke discussion', in S. Hedland (ed.) *Incentives and economic systems*, New York University Press, 315–45.

Krugman, P. R. (1986) 'Introduction: new thinking about trade policy', in P. R. Krugman (ed.) *Strategic trade policy and the new international economics*, MIT Press, Cambridge, Mass.

Kuznets, S. (1966) *Modern economic growth: rate, structure and spread*, Yale University Press, New Haven.

Kuznets, S. (1971) *Economic growth of nations: total output and production structure*, Belknap Press, Cambridge, Mass.

Laidler, D. (1981) 'Monetarism: an interpretation and an assessment', *Economic Journal*, 91, 1–28.

Lall, S. (1981) *Developing countries in the international economy: selected papers*, Macmillan, London.

Lall, S. (1983) *The new multinationals. The spread of Third World enterprises*, Wiley, Chichester.

Lampe, D. (ed.) (1988) *The Massachusetts miracle. High technology and economic revitalization*, MIT Press, Cambridge, Mass.

Lash, S. and J. Urry (1987) *The end of organised capitalism*, Polity Press, Cambridge.

Ledgerwood, G. (1985) *Urban innovation. The transformation of London's docklands, 1968–84*, Gower, Aldershot.

Leontief, W. W. (1951) *The structure of the American economy, 1919–1939. An empirical application of equilibrium analysis*, Oxford University Press, New York.

Leontief, W. W. (1953) 'Domestic production and foreign trade; the American capital position re-examined', *Proceedings*, American Philosophical Society, 97, 332–49.

Leontief, W. W. (1956) 'Factor proportions and the structure of American trade: further theoretical and empirical analysis', *Review of Economics and Statistics*, 38, 386–407.

Leontief, W. W., H. B. Chenery, P. G. Clark, J. S. Duesenberry, A. R. Ferguson, A. P. Grosse, R. N. Grosse, M. Holzman, W. Isard and H. Kistin (1953) *Studies in the structure of the American economy. Theoretical and empirical explorations in input-output analysis*, Oxford University Press, New York.

Lever, W. (1974) 'Manufacturing linkages and the search for suppliers and markets', in F. E. I. Hamilton (ed.) *Spatial perspectives on industrial organization and decision-making*, Wiley, London, 309–33.

Lever, W. F. (1989) 'International comparative aspects of government intervention in local economic policy', in D. Gibbs (ed.) 209–31.

Lever, W. F. and C. Moore (eds) (1986) *The city in transition. Policies and agencies for the economic regeneration of Clydeside*, Clarendon Press, Oxford.

Linder, S. B. (1961) *An essay on trade and transformation*, Wiley, New York.

Linge, G. J. R. and F. E. I. Hamilton (1981) 'International industrial systems', in F. E. I. Hamilton and G. J. R. Linge (eds) *Spatial analysis. Industry and the industrial environment*, vol. 2, *International industrial systems*, Wiley, Chichester, 1–117.

Lipietz, A. (1987) *Mirages and miracles. The crises of global Fordism*, Verso, London.

Lipsey, R. G. (1968) 'Structural and deficient-demand unemployment reconsidered', in B. J. McCormick and E. O. Smith (eds) *The labour market. Selected readings*, Penguin, Harmondsworth, 245–65.

Lloyd, P. E. and C. M. Mason (1984) 'Spatial variations in new firm formation in the United Kingdom: comparative evidence from Merseyside, Greater Manchester and South Hampshire' *Regional Studies*, 18, 207–20.

Lösch, A. (1954) *The economics of location*, Yale University Press, New Haven (first published in German 1944).

Luttrell, W. F. (1952) *The cost of industrial movement. A first report on the economics of establishing branch factories*, National Institute of Economic and Social Research, London, Occasional Paper XIV (published by Cambridge University Press, Cambridge).

Lutrell, W. F. (1962) *Factory location and industrial movement*, 2 vols, National Institute of Economic and Social Research, London.

Macgregor, B. D., R. J. Langridge, J. Adley and B. Chapman (1986) 'The development of high technology industry in the Newbury district', *Regional Studies*, 20, 433–48.

MacGregor, S. (1988) *The Poll Tax and the enterprise culture. The implication of recent local government legislation for democracy and the welfare state*, Centre for Local Economic Strategies, Manchester.

Mackay, R. R. (1982) 'Planning for balance; regional policy and regional employment – the United Kingdom experience', in W. Buhr and P. Friedrich *Planning under stagnation*, Nomos Verlagsgesellschaft, Baden-Baden, 57–94.

Malecki, E. J. (1986) 'Research and development and the geography of high-technology complexes', in J. Rees (ed.) *Technology, regions and policy*, Rowman & Littlefield, Totowa, 51–74.

Markusen, A. (1985) *Profit cycles, oligopoly and regional development*, MIT Press, Cambridge, Mass.

Markusen, A. (1987) *Regions. The economics and politics of territory*, Rowman & Littlefield, Totowa.

Markusen, A., P. Hall and A. Glasmeier (1986) *High tech America. The what, how, where, and why of the sunrise industries*, Allen & Unwin, Boston.

Marsden, D., T. Morris, P. Willman and S. Wood (1985) *The car industry. Labour relations and industrial adjustment*, Tavistock Publications, London.

Marshall, J. N., P. Wood, P. W. Daniels, A. McKinnon, J. Bachtler, P. Damesick, N. Thrift, A. Gillespie, A. Green and A. Leyshon (1988) *Services and uneven development*, Oxford University Press, Oxford.

Marshall, M. (1987) *Long waves of regional development*, Macmillan, London.

Marshall, N. (1987) 'Industrial change, linkages and regional development', in W. F. Lever (ed.) *Industrial change in the United Kingdom*, Longman, London, 108–22.

Martin, R. L. (1984) 'Redundancies, labour turnover and employment contraction in the recession: a regional analysis', *Regional Studies*, 18, 445–58.

Martin, R. L. (1985) 'Monetarism masquerading as regional policy? The government's new system of regional aid', *Regional Studies*, 19, 379–88.

Martin, R. L. (1986) 'Industrial restructuring, labour shake-out and the geography of recession', in M. Danson (ed.) *Redundancy and recession. Restructuring the regions*, Geo Books, Norwich, 1–22.

Martin, R. L. (1988a) 'The political economy of Britain's north-south divide', *Transactions*, Institute of British Geographers, 13, 389–418.

Martin, R. L. (1988b) 'De-industrialisation and state intervention: Keynesianism, Thatcherism and regions', in J. Mohan (ed.) *The political geography of contemporary Britain*, Macmillan, London, 87–112.

Martin, R. L. (1989) 'Regional imbalance as consequence and constraint in national economic renewal', in F. Green (ed.) *The restructuring of the UK economy*, Harvester Wheatsheaf, Hemel Hempstead, 80–97.

Martin, R. L. and J. S. C. Hodge (1983) 'The reconstruction of British regional policy: 2. Towards a new agenda', *Environment and Planning*, C, 1, 317–40.

Mason, C. M. and R. T. Harrison, (1986) 'The regional impact of public policy towards small firms in the United Kingdom', in D. E. Keeble and E. Wever (eds) *New firms and regional development in Europe*, Croom Helm, London, 224–55.

Mason, C. M. and R. T., Harrison (1989) 'Small firms policy and the "north-south" divide in the United Kingdom: the case of the Business Expansion Scheme', *Transactions*, Institute of British Geographers, 14, 37–58.

Massey, D. B. (1979) 'In what sense a regional problem?', *Regional Studies*, 13, 233–43.

Massey, D. B. (1984) *Spatial divisions of labour. Social structures and the geography of production*, Macmillan, London.

Massey, D. B. and J. Allen (eds) (1988) *Restructuring Britain. Uneven re-development: cities and regions in transition*, Hodder & Stoughton, London.

Massey, D. B. and R. A. Meegan (1978) 'Industrial restructuring versus the cities', *Urban Studies*, 15, 273–88.

Matthews, R. C. O. (1971) 'The role of demand management', in A. Cairncross (ed.) *Britain's economic prospects reconsidered*, Allen & Unwin, London, 13–35.

Matthews, R. C. O., C. H. Feinstein and J. C. Odling-Smee (1982) *British economic growth 1856–1973*, Clarendon Press, Oxford.

Mawson, J. and D. Miller (1983) *Agencies in regional and local development*, Occasional Paper 6, Centre for Urban and Regional Studies, University of Birmingham, Birmingham.

McCombie, J. S. L. (1988a) 'A synoptic view of regional growth and unemployment: I – the neoclassical theory', *Urban Studies*, 25, 276–81.

McCombie, J. S. L. (1988b) 'A synoptic view of regional growth and unemployment: II – the post-Keynesian theory', *Urban Studies*, 25, 399–417.

McKenzie, R. B. (1984) *Fugitive industry. The economics and politics of de-industrialization*, Ballinger, Cambridge, Mass.

Meade, J. E. (1982) *Stagflation*, vol. 1, *Wage-fixing*, Allen & Unwin, London.

Meltzer, A. H. (1989) *Keynes's monetary theory: a different interpretation*, Cambridge University Press, Cambridge.

Mensch, G. (1979) *Stalemate in technology. Innovations overcome the depression*, Ballinger, Cambridge, Mass. (first published in German 1975).

Miller, R. and M. Côté (1987) *Growing the next Silicon Valley. A guide for successful regional planning*, Lexington Books, Lexington.

Milne, S. S. (1989) '*New forms of manufacturing and their spatial implications. The case of the UK electronic consumer goods, high fidelity audio and domestic electrical appliance industries*', unpublished Ph.D. thesis, University of Cambridge, Cambridge.

Minford, P., D. Davies, M. Peel and A. Sprague (1983) *Unemployment. Cause and cure*, Martin Robertson, Oxford.

Monck, C. S. P., R. B. Porter, P. Quintas, D. J. Storey and P. Wynarczyk (1988) *Science parks and the growth of high technology firms*, Croom Helm, London.

Moore, B. and J. Rhodes (1982) 'A second great depression in the UK regions: can anything be done?', *Regional Studies*, 16, 323–33.

Moore, B., J. Rhodes and P. Tyler (1983) *The effects of government regional economic policy*, Discussion Paper, Department of Land Economy, University of Cambridge, Cambridge (published by HMSO, London, 1986).

Morgan, K. and A. Sayer, (1983) 'Regional inequality and the state in Britain', in J. Anderson, S. Duncan and R. Hudson (eds) Redundant spaces in cities and regions? Studies in industrial decline and social change, Academic Press, London, 17–49.

Moriarty, B. M. (1986) 'Productivity, industrial restructuring, and the deglomeration of American manufacturing', in J. Rees (ed.) *Technology, regions and policy*, Rowan & Littlefield, Totowa, 141–70.

Morishima, M. (1973) *Marx's economics. A dual theory of value and growth*, Cambridge University Press, Cambridge.

Morison, H. (1987) *The regeneration of local economies*, Clarendon Press, Oxford.

Morris, J. (1989) 'Japanese inward investment and the "importation" of sub-contracting complexes: three case studies', *Area*, 21, 269–77.

Moseley, M. J. (1974) *Growth centres in spatial planning*, Pergamon Press, Oxford.

Moseley, M. J. and P. M. Townroe (1973) 'Linkage adjustment following industrial movement', *Tijdschrift voor Economische en Sociale Geografie*, 64, 137–44.

Muegge, H. and W. B. Stöhr (eds) (1987) *International economic restructuring and the regional community*, Avebury, Aldershot.

Mueser, P. (1989) 'The spatial structure of migration: an analysis of flows between States in the USA over three decades', *Regional Studies*, 23, 185–200.

Myrdal, G. (1957) *Economic theory and under-developed regions*, Duckworth, London.

Nabseth, L. and G. F. Ray (eds) (1974) *The diffusion of new industrial processes. An international study*, Cambridge University Press, Cambridge.

Nicol, W. R. (1982) 'Estimating the effects of regional policy: a critique of the European experience', *Regional Studies*, 16, 199–210.

Nijkamp, P., R. van der Mark and T. Alsters (1988) 'Evaluation of regional incubator profiles for small and medium sized enterprises', *Regional Studies*, 22, 95–106.

North, D. C. (1955) 'Location theory and regional economic growth', *Journal of Political Economy*, 63, 243–58.

Nurske, R. (1961) *Patterns of trade and development*, Blackwell, Oxford.

Oakey, R. P. (1981) *High technology industry and industrial location: the instruments industry example*, Gower, Aldershot.

Oakey, R. P. (1985) 'High technology industries and agglomeration economies', in P. Hall and A. Markusen *Silicon landscapes*, Allen & Unwin, Boston, 94–115.

Oakey, R. P., R. Rothwell and S. Cooper (1988) *Management of innovation in high technology small firms: innovation and regional development in Britain and the United States*, Frances Pinter, London.

Oakey, R. P., A. T. Thwaites and P. A. Nash (1982) 'Technological change and regional development: some evidence on regional variations in product and process innovation', *Environment and Planning*, A, 14, 1073–86.

Ohlin, B. (1933) *Interregional and international trade*, Harvard University Press, Cambridge, Mass.

Organisation for Economic Co-operation and Development (OECD) (1989) *OECD employment outlook*, OECD, Paris.

Owen, D. W., M. G. Coombes and A. E. Gillespie (1986) 'The urban-rural shift and employment change in Britain, 1971–81', in M. Danson (ed.) *Redundancy and recession. Restructuring the regions?*, Geo Books, Norwich, 23–47.

Pasinetti, L. L. (1977) *Lectures in the theory of production*, Macmillan, London.

Patel, P. and L. Soete (1987) 'Technological trends and employment in the UK manufacturing sectors', in C. Freeman and L. Soete (eds) *Technical change and full employment*, Blackwell, Oxford, 122–68.

Peck, F. and A. Townsend (1987) 'The impact of technological change upon the spatial pattern of UK employment within major corporations', *Regional Studies*, 21, 225–39.

Perloff, H. S., E. S. Dunn, E. E. Lampard and R. F. Muth (1960) *Regions, resources and economic growth*, Johns Hopkins Press, Baltimore.

Perroux, F. (1950) 'Economic space, theory and applications', *Quarterly Journal of Economics*, 64, 89–104.

Perroux, F. (1955) 'Note sur la notion des "pôles de croissance"', *Économie appliquée*, 1 and 2, 307–20.

Perry, D. C. and A. J. Watkins (eds) (1977) *The rise of the sunbelt cities*, Sage, Beverly Hills.

Perry, M. and B. Chalkley (1985) 'New small factories in the public sector', *Area*, 17, 185–91.

Pfouts, R. W. (1960) *The techniques of urban economic analysis*, Chandler-Davis, West Trenton.

Phelps, E. S. (1967) 'Phillips curves, expectations of inflation and optimal unemployment over time', *Economica*, 34, 254–81.

Phillips, A. W. (1958) 'The relation between unemployment and the rate of change of money wage rates in the United Kingdom, 1861–1957', *Economica*, XXV, 283–99.

Piore, M. J. and C. F. Sabel (1984) *The second industrial divide. Possibilities for prosperity*, Basic Books, New York.

Pollard, S. (1982) *The wasting of the British economy*, Croom Helm, London.

Pollard, S. (1983) *The development of the British economy*, 3rd edition (1914–1980), Arnold, London.

Posner, M. V. (1961) 'International trade and technical change', *Oxford Economic Papers*, 13, 323–41.

Prais, S. J. (1981) *Productivity and industrial structure. A statistical study of manufacturing industry in Britain, Germany and the United States*, Cambridge University Press, Cambridge.

Pratten, C. F. (1985) *Applied macroeconomics*, Oxford University Press, Oxford.

Presbisch, R. (1951) *Economic survey of Latin America – 1949*, United Nations, New York.

Pred, A. R. (1966) *The spatial dynamics of US urban-industrial growth, 1800–1914: interpretive and theoretical essays*, MIT Press, Cambridge, Mass.

Pred, A. R. (1977) *City-systems in advanced economies. Past growth, present processes and future development options*, Hutchinson, London.

Pred, A. R. and G. E. Törnqvist (1973) *Systems of cities and information flows*, Lund Studies in Geography, Ser. B, 38, Royal University of Lund, Lund.

Pullen, M. J. and J. L. R. Proops (1983) 'The North Staffordshire regional economy: an input-output assessment', *Regional Studies*, 17, 191–200.

Rees, J. (ed.) (1986) *Technology, regions and policy*, Rowman & Littlefield, Totowa.

Rees, J. (1986) 'Introduction', in J. Rees (ed.) *Technology, regions and policy*, Rowman & Littlefield, Totowa, 1–20.

Rees, J., R. Briggs and D. Hicks (1985) 'New technology in the United States' machinery industry: trends and implications', in A. T. Thwaites and R. P. Oakey (eds) *The regional economic impact of technological change*, Frances Pinter, London, 164–94.

Rees, J., R. Briggs and R. Oakey (1986) 'The adoption of new technology in the American machinery industry', in J. Rees (ed.) *Technology, regions and policy*, Rowman & Littlefield, Totowa, 187–217.

Regional Studies Association (1983) *Report on an inquiry into regional problems in the United Kingdom*, Geo Books, Norwich.

Richardson, H. W. (1969) *Regional economics. Location theory, urban structure and regional change*, Weidenfeld & Nicolson, London.

Richardson, H. W. (1973) *Regional growth theory*, Macmillan, London.

Richardson, H. W. (1976) 'Growth pole spillovers: the dynamics of backwash and spread', *Regional Studies*, 10, 1–9.

Richardson, H. W. (1978) *Regional economics*, University of Illinois Press, Urbana.

Richardson, H. W. (1984) 'Approaches to regional development theory in Western-market economies', in G. Demko (ed.) *Regional development problems and policies in Eastern and Western Europe*, Croom Helm, London, 4–33.

Richardson, H. W. (1985) 'Input-output and economic base multipliers: looking backward and forward', *Journal of Regional Science*, 25, 607–61.

Rietveld, P. (1982) 'A general overview of multiregional economic models', in B. Issaev, P. Nijkamp, P. Rietveld and F. Snickars (eds) *Multiregional economic modelling: practice and prospect*, North-Holland, Amsterdam, 15–33.

Robinson, E. A. G. (1931) *The structure of competitive industry*, Cambridge University Press, Cambridge.

Robinson, E. A. G. (ed.) (1960) *The economic consequences of the size of nations*, Macmillan, London.

Robinson, J. (1933) *The economics of imperfect competition*, Macmillan, London (second edition published 1969).

Robson, B. T. (1973) *Urban growth. An approach*. Methuen, London.

Rodwin, L. (1963) 'Choosing regions for development', in C. J. Friedrich and S. E. Harris (eds) *Public policy: Yearbook of the Harvard University Graduate School of Public Administration*, 12, Harvard University Press, Cambridge, Mass., 132–46.

Rogerson, P. A. (1984) 'New directions in the modelling of interregional migration', *Economic Geography*, 60, 11–21.

Rostow, W. W. (1960) *The stages of economic growth: a non-Communist manifesto*, Cambridge University Press, Cambridge.

Rostow, W. W. (1978) *The world economy; history and prospect*, Macmillan, London.

Rothwell, R. (1982) 'The role of technology in industrial change: implications for regional policy', *Regional Studies*, 16, 361–9.

Rothwell, R. and W. Zegveld (1982) *Innovation and the small and medium sized firm. Their role in employment and in economic change*, Frances Pinter, London.

Rothwell, R. and W. Zegveld (1985) *Reindustrialization and technology*, Longman, Harlow.

Rowthorn, R. E. and J. R. Wells (1987) *De-industrialization and foreign trade*, Cambridge University Press, Cambridge.

Royal Commission on the Distribution of the Industrial Population (1940) *Report*, Cmd. 6153, HMSO, London.

Royal Geographical Society (1938) 'Memorandum on the geographical factors relevant to the location of industry', *Geographical Journal*, XCII, 499–526.

Samuelson, P. A. (1948) 'International trade and equalisation of factor prices', *Economic Journal*, 58, 163–84.

Samuelson, P. A. (1949) 'International factor-price equalisation once again', *Economic Journal*, 59, 181–96.

Samuelson, P. A. and W. D. Nordhaus (1985) *Economics*, McGraw-Hill, New York, 12th edition.

Saxenian, A. (1985) 'The genesis of Silicon Valley', in P. Hall and A. Markusen (eds) *Silicon landscapes*, Allen & Unwin, Boston, 20–34.

Schamp, E. W. (1987) 'Technology parks and interregional competition in the Federal Republic of Germany', in B. van der Knapp and E. Wever (eds) *New technology and regional development*, Croom Helm, London, 119–35.

Schoenberger, E. (1986) 'Competition, competitive strategy, and industrial change: the case of electronic components', *Economic Geography*, 62, 321–33.

Schoenberger, E. (1987) 'Technological and organizational change in automobile production: spatial implications', *Regional Studies*, 21, 199–214.

Schoolman, M. (1986) 'Solving the dilemma of statesmanship. Reindustrialization through an evolving democratic plan', in M. Schoolman and A. Magid (eds) *Reindustrializing New York State. Strategies implications, challenges*, State University of New York Press, New York, 3–49.

Schoolman, M. and A. Magid (eds) (1986) *Reindustrializing New York State. Strategies, implications, challenges*, State University of New York Press, New York.

Schultz, T. W. (1961) 'Investment in human capital', *American Economic Review*, 51, 1–17.

Schultz, T. W. (1981) *Investing in people. The economics of population quality*, University of California Press, Berkeley.

Schumpeter, J. A. (1939) *Business cycles. A theoretical, historical, and statistical analysis of the capitalist process*, McGraw-Hill, New York.
Schumpeter, J. A. (1943) *Capitalism, socialism and democracy*, Allen & Unwin, London.
Schumpeter, J. A. (1954) *History of economic analysis*, Allen & Unwin, London.
Scott, A. J. (1986) 'High technology industry and territorial development: the rise of the Orange County complex, 1955–1984', *Urban Geography*, 7, 3–45.
Scott, A. J. (1988a) 'Flexible production systems and regional development: the rise of new industrial spaces in North America and Western Europe'. *International Journal of Urban and Regional Research*, 12, 171–86.
Scott, A. J. (1988b) *New industrial spaces. Flexible production organization and regional development in North America and Western Europe*, Pion, London.
Scott, A. J. and E. C. Kwok (1989) 'Inter-firm subcontracting and locational agglomeration: a case study of the printed circuits board industry in Southern California', *Regional Studies*, 23, 405–16.
Scott, A. J. and M. Storper (eds) (1986) *Production, work and territory*, Allen & Unwin, London.
Scott, A. J. and M. Storper (1987) 'High technology industry and regional development: a theoretical critique and reconstruction', *International Social Science Journal*, 112, 215–32.
Scottish Development Agency (1989) *Annual Report. Nineteen eighty nine*, SDA, Glasgow.
Segal, N. S. (1985) 'The Cambridge phenomenon', *Regional Studies*, 19, 563–70.
Segal, Quince, Wicksteed (1985) *The Cambridge phenomenon: the growth of high technology industry in a university town*, Segal, Quince, Wicksteed, Cambridge.
Select Committee on Oveseas Trade (1985) *Report*, House of Lords Session 1984–85, HMSO, London.
Sellgren, J. (1989) 'Assisting local economies: an assessment of emerging patterns of local authority economic development activities', in D. C. Gibbs (ed.) *Government policy and industrial change*, Routledge, London, 232–60.
Semmler, W. (1984) *Competition, monopoly, and differential profit rates*, Columbia University Press, New York.
Sheppard, E. and T. J. Barnes (1986) 'Instabilities in the geography of capitalist production: collective vs. individual profit maximization', *Annals*, Association of American Geographers, 76, 493–507.
Sheppard, E. and T. J. Barnes (1990) *The capitalist space economy: geographical analysis after Marx, Sraffa and Ricardo*, Unwin Hyman, London.
Shutt, J. and R. C. Whittington (1987) 'Fragmentation strategies and the rise of small units: cases from the North West', *Regional Studies*, 21, 13–23.
Smith, N. (1984) *Uneven development. Nature, capital and the production of space*, Blackwell, London.
Smith, W. (1949) *An economic geography of Great Britain*, Methuen, London.
Solomon, E. (ed.) (1984) *International patterns of inflation: a study in contrasts*, American Council of Life Insurance and The Conference Board.
Solomon, S. (1988) *Phases of economic growth, 1850–1973. Kondratieff waves and Kuznets swings*, Cambridge University Press, Cambridge.
Southwick, L. (1986) 'Local economic development and the state', in M. Schoolman and A. Magid (eds) Reindustrializing New York State. Strategies, implications, challenges, State University of New York Press, Albany, 147–64.
Spencer, K., A. Taylor, B. Smith, J. Mawson, N. Flynn and R. Batley (1986) *Crisis in the industrial heartland. A study of the West Midlands*, Clarendon Press, Oxford.
Steele, D. B. (1969) 'Regional multipliers in Great Britain', *Oxford Economic Papers*, 21, 268–92.

Steele, D. B. (1972) 'A numbers game (or the return of regional multipliers)', *Regional Studies*, 6, 115–30.

Steiner, M. (1985) 'Old industrial areas: a theoretical approach', *Urban Studies*, 22, 387–98.

Stigler, G. J. (1965) *Essays in the history of economics*, University of Chicago Press, Chicago.

Stillwell, J. C. H. (1978) 'Interzonal migration: some historical tests of spatial inter-action models', *Environment and Planning*, A, 10, 1187–1200.

Stöhr, W. and D. R. F. Taylor (eds) (1981) *Development from above or below? The dialectics of regional planning in developing countries*, Wiley, Chichester.

Stoney, P. J. M. (1986) 'The employment impact of the Merseyside motor-vehicle assembly industry', in M. Danson (ed.) *Redundancy and recession. Restructuring the regions?*, Geo Books, Norwich, 65–81.

Storey, D. J. (ed.) (1985) *Small firms in regional economic development. Britain, Ireland and the United States*, Cambridge University Press, Cambridge.

Storey, D. J. (1987) *The performance of small firms. Profits, jobs and failures*, Croom Helm, London.

Strange, S. (1988) *States and markets. An introduction to international political economy*, Frances Pinter, London.

Taylor, E. G. R., G. H. J. Daysh, H. J. Fleure and W. Smith (1938) 'Discussion on the geographical distribution of industry', *Geographical Journal*, XCII, 22–32.

Taylor, M. J. (1973) 'Local linkage, external economies and the ironfoundry industry of the West Midlands and East Lancashire conurbations', *Regional Studies*, 7, 387–400.

Taylor, M. J. and N. J. Thrift (1982) 'Introduction', in M. J. Taylor and N. J. Thrift (eds) *The geography of multinationals. Studies of the spatial development and economic consequences of multinational corporations*, Croom Helm, London, 1–13.

Taylor, M. J. and N. J. Thrift (eds) (1986) *Multinationals and the restructuring of the world economy*, Croom Helm, London.

Thirlwall, A. P. (1980) 'Regional problems are "balance-of-payments" problems', *Regional Studies*, 14, 419–25.

Thomas, B. (1954) *Migration and economic growth. A study of Great Britain and the Atlantic economy*, Cambridge University Press, Cambridge.

Thomas, D. (1983) 'Shipbuilding – demand linkage and industrial decline', in K. Williams, J. Williams and D. Thomas *Why are the British bad at manufacturing?*, Routledge, London, 179–216.

Thomas, M. D. (1985) 'Regional economic development and the role of innovation and technological change', in A. T. Thwaites and R. P. Oakey (eds) *The regional economic impact of technological change*, Frances Pinter, London, 13–35.

Thompson, C. (1989) 'High-technology theories and public policy', *Environment and Planning*, C, 7, 121–52.

Thompson, W. R. (1965) *A preface to urban economics*, Johns Hopkins, Baltimore.

Thwaites, A. T., A. Edwards and D. C. Gibbs (1982) *Interregional diffusion of pro-duction innovations in Great Britain*, Final Report, Centre for Urban and Regional Development Studies, University of Newcastle upon Tyne, Newcastle.

Thwaites, A. T. and R. P. Oakey (eds) (1985) *The regional economic impact of technolo-gical change*, Frances Pinter, London.

Tiebout, C. M. (1956) 'A pure theory of local expenditure', *Journal of Political Economy*, 64, 416–24.

Todd, D. (1985) *The world shipbuilding industry*, Croom Helm, London.

Treasury Officials (1985) *The relationship between employment and wages. Empirical evidence for the United Kingdom*, HM Treasury, London.

Ture, N. B. (1982) 'The economic effects of tax changes in neoclassical analysis', in R. H. Fink (ed.) *Supply-side economics: a critical appraisal,* University Publications of America, Washington, DC, 33–69.

Tyson, L. and J. Zysman (1983) 'American industry in international competition', in J. Zysman and L. Tyson (eds) *American industry in international competition. Government policies and corporate strategies,* Cornell University Press, Ithaca, 15–59.

US Bureau of the Census (1987) *Statistical abstract of the United States: 1988,* US Government Printing Office, Washington, DC.

Vane, H. and T. Caslin (1987) *Current controversies in economics,* Blackwell, Oxford.

Vanhove, N. and L. H. Klaassen (1987) *Regional policy: a European approach,* Avebury, Aldershot.

Vaughan, R. and R. Pollard (1986) 'State and federal policies for high-technology development', in J. Rees (ed.) Technology, regions, and policy, Rowman & Littlefield, Totowa, 268–81.

Vernon, R. (1966) 'International investment and international trade in the product cycle', *Quarterly Journal of Economics,* 80, 190–207.

Vernon, R. (1979) 'The product cycle hypothesis in a new international environment', *Oxford Bulletin of Economics and Statistics,* 41, 255–67.

Wadley, D. (1986) *Restructuring the regions. Analysis, policy model and prognosis,* Organisation for Economic Co-operation and Development, Paris.

Wannop, U. and R. Leclerc (1987) 'Urban renewal and the origins of GEAR', in D. Donnison and A. Middleton (eds) *Regenerating the inner city: Glasgow's experience,* Routledge & Kegan Paul, London, 61–71.

Weaver, C. (1984) *Regional development and the local community: planning, politics and social context,* Wiley, Chichester.

Weber, A. (1929) *Alfred Weber's theory of the location of industry,* University of Chicago Press, Chicago (first published in German 1909).

Webman, J. A. (1982) *Reviving the industrial city. The politics of urban renewal in Lyon and Birmingham,* Rutgers University Press, New Brunswick, NJ.

Weiner, M. (1981) *English culture and the decline of the industrial spirit 1850–1980,* Cambridge University Press, Cambridge.

Weiss, M. A. (1985) 'High technology industries and the future of employment', in P. Hall and A. Markusen (eds) *Silicon landscapes,* Allen & Unwin, Boston, 80–93.

Weitzman, M. L. (1984) *The share economy. Conquering stagflation,* Harvard University Press, Cambridge, Mass.

Wenger, G. C. (1980) *Mid-Wales: deprivation or development. A study of employment in selected communities,* Social Science Monograph 5, University of Wales Press, Cardiff.

Wheaton, W. C. (1979a) 'Metropolitan growth, unemployment, and interregional factor mobility', in W. C. Wheaton (ed.) *Interregional movements and regional growth,* The Urban Institute, Washington, DC, 237–53.

Wheaton, W. C. (1979b) 'Introduction', in W. C. Wheaton (ed.) *Interregional movements and regional growth,* The Urban Institute, Washington, DC, 1–12.

Whittington, R. C. (1984) 'Regional bias in new firm formation in the UK', *Regional Studies,* 18, 253–6.

Williams, K. (1983) 'BMC/BLMC/BL – A misunderstood failure', in K. Williams, J. Williams and D. Thomas *Why are the British bad at manufacturing?* Routledge, London, 217–81.

Williams, K., J. Williams and D. Thomas (1983) *Why are the British bad at manufacturing?,* Routledge, London.

Williamson, J. G. (1965) 'Regional inequalities and the process of national development', *Economic Development and Cultural Change*, 13, 1–84.

Williamson, J. G. (1983) *The open economy and the world economy. A textbook in international economics*, Basic Books, New York.

Willis, K. (1974) *Problems in migration analysis*, Saxon House, Farnborough.

Wilmers, P. and B. Bourdillon (eds) (1985) *Managing the local economy: planning for employment and economic development*, Geo Books, Norwich.

Wilson, A. G. (1971) 'A family of spatial interaction models, and associated developments', *Environment and Planning*, 3, 1–32.

Wilson, A. G. (1980) 'Comments on Alonso's "theory of movement"', *Environment and Planning*, A, 12, 727–32.

Wilson, T. (1968) 'The regional multiplier – a critique', *Oxford Economic Papers*, 20, 374–93.

Wise, M. J. (1949) 'On the evolution of the gun and jewellery quarters in Birmingham', *Transactions*, Institute of British Geographers, 15, 59–72.

Wood, P. A. (1986) 'The anatomy of job loss and creation: Some speculations on the role of the "producer services" sector', *Regional Studies*, 20, 37–46.

Woods, R. (1979) *Population analysis in geography*, Longman, London.

World Bank (1986) *World development report 1986*, Oxford University Press, New York.

Wright, D. M. (1962) *The Keynesian system*, Fordham University Press, New York.

Young, A. A. (1928) 'Increasing returns and economic progress', *Economic Journal*, XXXVIII, 527–42.

Youngson, A. J. (1967) *Overhead capital: a study in development economics*, Edinburgh University Press, Edinburgh.

Yuill, D. (ed.) (1982) *Regional development agencies in Europe. An international comparison of selected agencies*, Gower, Aldershot.

Yuill, D. and K. J. Allen (1980-) *European regional incentives*, Centre for the Study of Public Policy, University of Strathclyde, Glasgow. Annual issues.

Yuill, D., K. J. Allen and C. Hull (eds) (1980) Regional policy in the European community, Croom Helm, London.

Zipf, G. K. (1949) *Human behavior and the principle of least effort*, Addison-Wesley, Cambridge, Mass.

# Author Index

# Subject Index